Philanthropic Foundations and the Globalization of Scientific Medicine and Public Health

Proceedings of a conference jointly sponsored by Quinnipiac University and the Rockefeller Archive Center with additional support from the Dreyfus Health Foundation

Edited by
Benjamin B. Page
David A. Valone

University Press of America,® Inc.
Lanham · Boulder · New York · Toronto · Plymouth, UK

University Press of America,® Inc.
4501 Forbes Boulevard
Suite 200
Lanham, Maryland 20706
UPA Acquisitions Department (301) 459-3366

Estover Road
Plymouth PL6 7PY
United Kingdom

Printed in the United States of America
British Library Cataloging in Publication Information Available

Library of Congress Control Number: 2007926459
ISBN-13: 978-0-7618-3819-7 (paperback : alk. paper)
ISBN-10: 0-7618-3819-8 (paperback : alk. paper)

∞™ The paper used in this publication meets the minimum
requirements of American National Standard for Information
Sciences—Permanence of Paper for Printed Library Materials,
ANSI Z39.48—1984

Contents

List of Contributors

Marianne Fedunkiw completed her PhD at the University of Toronto in 2000. Her areas of research interest include women doctors and World War I, motion pictures as public health propaganda tools, and the influence of private philanthropy on medical education and research. Her publications include: *Rockefeller Foundation Funding and Medical Education in Toronto, Montreal and Halifax* (McGill-Queen's University Press, Montreal, 2005) and *Training Physicians in Alberta, for Alberta,* (University of Alberta Press, Edmonton, 2007) as well as "'German methods', 'unconditional gifts' and the full-time system: The case study of the University of Toronto, 1919-1923", *Canadian Bulletin of Medical History*, 21 (2004): 5-39, and "Malaria films as a public health tool, 1940-2002," *The American Journal of Public Health* 93 (July 2003): 1046-1056.

Peter C. Goldmark, Jr. currently directs the Climate and Air program for Environmental Defense. Prior to joining Environmental Defense, he was Chairman and CEO of the International Herald Tribune. He has had exceptional careers in both the public and private sectors. He served as president of the Rockefeller Foundation from 1988 to 1997, encouraging its involvement in environmental issues, particularly as they related to energy. He was also a trustee of the Rockefeller Brothers Fund, a member of Board Overseers and Chair of Harvard University's Finance Committee, director of Knight Ridder Inc. and of the Dreyfus Third Century Mutual Fund, a member of the National Commission on Civic Renewal and a trustee of the Financial Accounting Foundation. In addition, he serves as a board member of Lend Lease Corporation, is a trustee of the Whitehead Institute for Biomedical Research and a member of the Council on Foreign Affairs. He has taught courses at the JFK School of Government, Harvard; Yale College; The New School; Brandeis University and returns to Woodrow Wilson School, Princeton University as Visiting Professor of Public and International Affairs in spring 2007. He holds a BA from Harvard University.

Soma Hewa is the principal investigator of a research program on philanthropy, civil society, and non-profit activities based in Châteauguay, Québec, Canada. He taught sociology at Mount Royal College, Calgary, and the University of British Columbia, Vancouver. He has published widely in sociology of health, philanthropy, and sociological theory. His major publications include *Colonialism, Tropical Disease and Imperial Medicine: Rockefeller Philanthropy in Sri Lanka* (University Press of America, 1995), and *Globalization, Philanthropy* *and Civil Society: Toward a New Political Culture in the Twenty-First Century* (Springer, 2005, co-edited with Darwin H. Stapleton).

Shirish N. Kavadi is an independent researcher and consultant to non-governmental organizations and development agencies, based in Pune, India. He has degrees in History and Political Science from the University of Bombay (now Mumbai). Since 1980 his career has traversed both academia and the NGO sector. His area of research includes health policy, the politics of health and history of health/medicine. His specific interest is in philanthropy's interface with public health and medicine. He has published a study on the Rockefeller Foundation's public health program in colonial India and has contributed articles to academic journals and chapters to edited volumes. At present he is completing a study on philanthropy and scientific medicine that examines the role of the Tata family (who set up India's first philanthropic trusts and several national institutes) and the Rockefeller Foundation in institutionalizing medical research in India covering the period 1890 to 1960. Kavadi had been a recipient of research grants from the Rockefeller Archive Centre, the Wellcome Trust, London, and the Sir Dorabji Trust, Mumbai.

Stephen R. Latham, JD, PhD, is Professor of Law and Director of the Center on Health Law & Policy at Quinnipiac University School of Law, where he teaches bioethics and health law. He received his JD from Harvard Law School and his doctorate from the University of California at Berkeley. He is former Director of Ethics Standards for the American Medical Association, a former graduate fellow of Harvard's Safra Center on Ethics, and a former Research Fellow at the University of Edinburgh's Institute for Advanced Studies in the Humanities. His many writings on bioethics and medical professionalism have appeared, *inter alia*, in *JAMA*, *NEJM*, the *American Journal of Bioethics*, *the Journal of Law, Medicine and Ethics*, the *Blackwell Guide to Bioethics* and the *Cambridge History of Bioethics*. Latham currently serves on the board of the American Society for Bioethics and Humanities, and on Connecticut's Stem Cell Research Advisory Committee.

Dorothy Brewster-Lee is the HIV/AIDS Senior Technical Advisor at Catholic Relief Services. A physician with extensive experience in international health, Dr. Brewster-Lee has spent the past fifteen years with organizations providing AIDS prevention and care services around the world. As the Coordinator for International Health Ministries for the Presbyterian Church (USA), she was responsible for working with 28 mission hospitals and more than 40 community/congregation-based healthcare projects of partner churches in 34 countries. Now working for CRS, she participated in the organization's initial work to scale-up the response to orphans and vulnerable children in resource limited settings. She is currently leading their effort to increase HIV primary behavior change interventions in Africa. Dorothy Brewster-Lee received her Masters of Public Health from John Hopkins University (1981) and her Medical Degree from Howard University (1978).

Chinyelu Lee is currently a doctoral candidate completing a dissertation on *Democracy, Dependency and Development: Constraining Local Responses to HIV/AIDS in Malawi* at Yale University. Mr. Lee also teaches Political Theory and Urban Politics at Quinnipiac University.

Socrates Litsios retired from WHO in 1997 and currently lives in Baulmes, Switzerland (email: litsioss@bluewin.ch). His interest in medical history developed while he was working with the malaria action program in the 1980s. This led to the writing of the book *The Tomorrow of Malaria* and numerous articles (see www.litsios.com). He is currently working on the official history of WHO's 3rd decade (1968-1977) and editing *Plague Legends II: In search of public health (1830-1940)*, which follows *Plague Legends: from the miasmas of Hippocrates to the microbes of Pasteur* which was published in 2001.

Paola Mejia, MD, MPH, is a PhD candidate in the History of Public Health program at Columbia University. She is currently working on a history of malaria control in Colombia.

Benjamin B. Page earned his PhD in Social Philosophy at Florida State University in a program that required students to take roughly one third of their courses in the social sciences. As a result he was also able to earn a post doctoral MS in Urban and Regional Health Planning. His interest in the work of the Rockefeller Foundation in globalizing "scientific" medicine and public health emerged during a year (1975-6) spent in Czechoslovakia studying Biomedical Ethics in the context of a socialist economic system—and serendipitously discovering the role the RF had played in the development of the country's public health work between the world wars. Professor Page has continued to work in both Philosophy and Health Planning during the years he has taught at Quinnipiac University, and has published articles about the RF in Czechoslovakia in both English-language and Czech journals.

Dr. Thomas Rathwell is Professor and Director of the School of Health Services Administration at Dalhousie University and Head of the affiliated World Health Organization Collaborating Centre for Health Care Systems Research and Development. He is a board member of both the Nova Scotia Health Research Foundation and the Canadian Society for International Health. He is the Dalhousie University lead for the consortium of Atlantic Canada universities involved in the Atlantic Canada Regional Training Centre in Applied Health Services Research.

Ronald T. Rozett, MD, MPH is a graduate of Harvard Medical School and the Harvard School of Public Health. After clinical training, he became a member of the Internal Medicine faculty of the Yale University School of Medicine. Following several years in academic medicine, he practiced medicine as a primary care internist and geriatrician in a New Haven prepaid group practice. He also served as its Medical Director for nine years. After 17 years at the group practice, he moved to a rehabilitation hospital to pursue his interest in the management of chronic disease and disability and continue his work as an administrator. He is currently an Executive in Residence at the Quinnipiac University School of Business, the Director of its Master of Health Administration Program, and the Medical Director of the University's Physician Assistant Program.

Dr. Carl Taylor has dedicated his life to improving the health care of people throughout the world, building on the principle of equity. He was one of the founders and first director of the Department of International Health at JHSPH in 1961, the first department of its kind at any school of public health. Over the course of 30 years, he worked in India, first as director of Memorial Hospital, a Presbyterian Mission, and then as head of a preventive social medicine department at Christian Medical College in Ludhiana. Dr. Taylor served as UNICEF director for China from 1984 through 1987, and as primary WHO consultant in preparing documents for Alma Ata, a World Conference in 1978 on Primary Health Care for the first time. Born in India, the son of medical missionaries for the Reformed Presbyterian mission in the Himalayas, Taylor says he inherited his profession. His own training in medicine and later a DrPH, both at Harvard, carried on the tradition of involving his family in this work. His late wife, Mary, professor emeritus in education at Towson University, as well as their children, participated in Taylor's work, while the family lived in China and India.

Henry G. Taylor, MD, MPH teaches the practice of public health, conducting research on the critical workforce and organizations that help keep the public healthy. This focus builds on his seven years as State Health Officer and Public Health Commissioner for the State of West Virginia under three different Governors. Previously, he and his wife developed an innovative rural primary health care system, Pendleton Community Care. Dr. Taylor is a Board Certified Internist, who obtained his MD from Harvard in 1979 and his MPH from Johns Hopkins in 1996.

James Sensenig is a naturopathic physician in private practice in Hamden, Connecticut. He was the founding President of the American Association of Naturopathic Physicians and served as Chair of the Association's Government Affairs Committee for over ten years. He has been the Dean of the National College of Naturopathic Medicine in Portland, Oregon and the College of Naturopathic Medicine at the University of Bridgeport. Currently, he is a visiting professor of medical history and philosophy at both the University of Bridgeport and the Southwest College of Naturopathic Medicine in Tempe, Arizona.

David A. Valone is an assistant professor of history at Quinnipiac University, and holds degrees from Princeton University and the University of Chicago. He previously worked on the staff of the Institute Archives of the California Institute of Technology, where he processed the archival records of the Human Betterment Foundation. His areas of interest include nineteenth century British and Irish history as well as the history of science and medicine. His published works include *Ireland's Great Hunger: Silence, Memory and Commemoration* (University Press of America, 2002) co-edited with Christine Kinealy, as well as articles on the history of linguistics in the nineteenth century, on the history of *in-vitro* fertilization, and on educational and religious debates in Britain during the 1820s.

Acknowledgments

This book would not have been possible without the help and support of many individuals. Initial support for the project came from the Director of the Rockefeller Archive Center, Darwin Stapleton. His continued enthusiasm for the work has been critical throughout the entire project. Once the idea of a joint conference between the RAC and Quinnipiac University began to take shape, many more people became involved who deserve much of the credit for making the conference and this publication a success. At the Rockefeller Archive Center, Kenneth Rose was instrumental in helping to shape the intellectual vision of the conference. Norine Hochman put a tremendous amount of time and effort into organizing the myriad details of planning and hosting a conference. Her efforts made the conference possible and maintained the momentum for the publication of this volume. Without her this book would not have come to fruition.

At Quinnipiac, President John L. Lahey, Senior Vice-President for Academic Affairs Kathleen McCourt, and the Dean of Liberal Arts Hans Bergmann provided financial and institutional support for the Quinnipiac part of the conference. In the Quinnipiac School of Law, Robin Utley provided superb on-the-ground supervision of the Law School's audio-visual technology. Betsy Delaney made everything else at Quinnipiac possible with her usual warmth, good humor, and energy. Emily Weed provided invaluable assistance with editing and proofreading the final manuscript.

During the conference, Peter Goldmark gave a stirring keynote address, reproduced here. His energy and vision brought real life to the conference. All of the conference participants also contributed significantly to the success of the conference itself. Their patience and perseverance in sticking with the book project over several long years has been absolutely critical. Quinnipiac students attended the conference in great numbers and provided interesting perspectives and insightful commentary on many of the papers.

Finally, Patti Belcher and Catherine Forrest Getzie at University Press of America ushered the work through the editorial and publication process and deserve considerable credit for their patient and efficient handling of the manuscript.

Introduction

Benjamin B. Page and David A. Valone

The conference from which the following papers are selected emerged from discussions between an individual researcher (Benjamin Page) at the Rockefeller Archive Center (RAC) and the Center's Executive Director, Darwin Stapleton. Following these initial conversations, a small *ad hoc* team was formed at the researcher's home institution—Quinnipiac University, in Hamden, Connecticut—to collaborate with the RAC in developing the project.

From the outset we worked from a few shared starting points. We wanted a conference that would produce a publication that would be useful to foundations and other agencies in international health. For this purpose, interdisciplinary, historical, and contemporary viewpoints, including perspectives from recipient countries, would all be important. It was decided that it should be a residential conference, with participants staying at the same location in the hope of providing opportunity and "safe space" for the informal contacts and discussions that can often be even more valuable than the formal parts of a conference. Finally, we wanted to involve people from the target foundations and agencies in planning the event and the themes to be explored. Two pre-conference planning sessions were held with representatives from foundations, the Pan-American Health Organization, and the World Health Organization.

What emerged after all of these discussions was a two and a half day event jointly sponsored by the Rockefeller Archive Center and Quinnipiac University, with generous supplemental support from the Dreyfus Health Foundation. In addition, the RAC offered stipends for participants using its facilities in connection with their presentations. The invitation indicated the basic premise: "throughout much of the past century philanthropic foundations led the way in defining and responding to the health-related issues that have affected large parts of the world, efforts that have included playing a major role in the creation of the health organizations of the League of Nations and the United Nations. Although this has benefited many peoples and nations, there is need today to reflect on the experience of this work and on the insights, questions, and warnings such reflection may offer to the future." The emphasis was to be on foundation work done since around 1980, the beginning of the large-scale shift toward privatization and market economics and the end of the Cold War, but with historical grounding going back to the earliest international work in health and medicine, initiated by the Rockefeller Foundation in 1913. On this basis, the main themes to which invitees were requested to contribute were:

- "The Changing Scene"—the most needed roles foundations might play in a changed and changing global environment;
- "Agenda-Setting Diseases"—the nature and effects of, and responses to, the key diseases of the present: AIDS, SARS, tuberculosis;
- "Goals and Values"—possible tensions among key values (health, equity, liberty, economic development) and interests (local, international, commercial, health-related);
- "Evaluation and Accountability"—should foundations and non-governmental organizations in international health be responsible to any outside agency and if so, for what?

Following a welcome reception and dinner at the conference hotel, the participants viewed an early educational film from the anti-hookworm campaign of the Rockefeller Sanitary Commission—precursor of the Foundation's International Health Commission (Board, Division). The formal part of the conference began with comments by Dr. Timothy Evans, who had recently moved from the Commonwealth Foundation to the Geneva office of the World Health Organization, and Prof. Stanley Katz, who described the allied health training program Quinnipiac had initiated—and the fall of the Shah prematurely terminated—in Iran. The first day's sessions, open to Quinnipiac students and the interested public, took place at the Quinnipiac University School of Law. The second day, a working day open only to invited participants, was held at the RAC, in Tarrytown, New York; it included a tour of the Center, additional formal paper sessions, and a facilitated discussion among the participants. Although formal evaluations were not requested, perhaps the most frequently expressed comment was appreciation for the diversity of backgrounds and professional expertise that had been gathered. We trust that something of that will also be found of value to readers of this collection of papers.

The order of the papers in this volume attempts to capture something of the spirit of the conference by opening with a consideration of the historical and philosophical foundations of the involvement of philanthropies in the internationalization of western "scientific" medicine and public health, followed by examinations of a series of related contemporary issues. The collection begins with philosophically informed examination by Benjamin Page of the early development of the Rockefeller Foundation's (RF's) involvement in international health. From the first development of an International Health Board to the creation of a vision for medical education on a world wide scale, Page's essay explores the early successes the RF achieved in planting the seeds of scientific medicine around the world. At the same time, however, the essay calls into question both the intellectual foundations of these efforts and their long term sustainability.

Marianne P. Fedunkiw's study of the RF's Malaria campaigns raises questions about the role of the RF in creating and disseminating information about its health activities to a public often ill equipped to understand the scientific foundations of its work. In the RF's 1925 *Malaria* film, Fedunkiw shows, Rockefeller officials were acutely aware of what they perceived as limitations of the "native or ignorant mind" (p. 18). Film was viewed as a particularly effective media for communicating with those not fully capable of understanding the scientific and medical subtleties of disease prevention and eradication.

David A. Valone discusses similar "limitations" in connection with his study of the role of philanthropic foundations in the emergence of the world population control movement after World War Two. Valone argues that while prior to the War some foundations (including the RF) actively supported eugenic sterilization, in the decade of the 1940s the focus of these efforts shifted from eugenics to population control. Because many of the same individuals and institutions were involved, there would seem to be clear parallels in the ideological and political foundations of these two movements; his paper examines the continuing influence that eugenic ideology had over the issue of population control well into the late 20th century.

In his examination of the early history of the RF malaria control efforts, Socrates Litsios focuses on a key fissure within RF officials concerning the degree to which malaria control must take into account broader social aspects of disease control. In his detailed study of the contrasting approaches of Selskar Gunn and Paul Russell, Litsios raises several critical issues about the ways in which differences in individual background and training, personal experiences, and social views had a profound impact on the approaches that might be advocated concerning public health. His paper raises deep questions about varying definitions of disease, public health, and the impact that these varied views had upon decisions made by philanthropic organizations, and by implication governments and NGOs, in pursuing health around the world.

Public health considerations are also at the heart of Soma Hewa investigation of the experience of the RF's Health Unit Program in Sri Lanka. Following its success in the treatment of hookworm in the U.S., the RF embarked on similar campaigns throughout what was found to be a global "hookworm belt" that included Sri Lanka. Working on the assumption that preventable diseases endemic to a particular area should be tackled first, the RF created health units throughout the region to undertake projects in sanitation, vaccination, and health education. In this case, Hewa argues, the RF's programs successfully laid the foundation for an extensive public health care system that today gives access to basic health care for an overwhelming part of the Sri Lankan population.

Focusing on the RF's leadership in the battle against Yellow Fever, Paola Mejia scrutinizes the origins and consequences of the partnership established during the 1930s between the government of Colombia and the RF. Rodriguez shows how the Colombia government enthusiastically participated in yellow fever control efforts during the nationalistic surge following the conflict with Peru. Over time, however, the tensions between the basic science focus of the RF and the pressing public health needs of the Colombia population led to a souring of the relationship, culminating in the government's abrupt termination of the partnership in 1948.

Mejia argues that Colombia paid a high price to support RF research that in the end benefited the nation's population only marginally.

In the last essay in this historical section, Shirish N. Kavadi examines the difficult relationships that existed between the RF and the government in India during the interwar period. He argues that by tying its efforts so closely into the governments of the nations in which it was operating, the RF frequently was forced to change its goals or even compromise its principles as it struggled with the complex structure and political wrangling within particular nations. He focuses his examination on the history of the RF's operations in India during the 1930s, when political uncertainties undermined the creation of a RF funded public health program in Madras.

The second half of the book considers the contemporary state of affairs regarding the work of philanthropic foundations and the delivery of health services globally. In their paper on "Community Based Primary Health Care," Carl and Henry Taylor examine three case studies: Tibet, the Indian Himalayas, and Peru. Their experiences make clear the difficulties faced around the globe in creating local health care systems. They stress the need for the development of a true partnership between community based efforts by local people to improve their own health care and the demands of governments and other organizations that work primarily with a top-down approach to the issue of public health. In their view, it is particularly important to avoid the tendency to crush the self-reliance and local empowerment of communities as large scale systematic changes are brought on by economic development, population growth, and political centralization.

James Sensenig offers another interesting perspective on this dichotomy by considering the role of indigenous or alternative medical theories and practices in the context of the globalization of western medicine. He argues that in forcing a single model upon the world, the West has lost sight of both the power of alternative understandings of health and the potential therapeutic and cost benefits they might hold. By allowing only its own model to hold legitimacy, Sensenig argues that the West has lost a considerable amount with its rigorously reductionist and materialist medicine; he urges the value of foundation support for "biodiversity" among competing models of medicine.

In his essay on "Agenda Setting Diseases," Ronald Rozett considers the role of diseases that have such a deep impact on a society that they force the hand of political, medical, and economic decision makers in the places that they affect. According to his analysis, the diseases that most clearly fit this categorization are AIDS, tuberculosis, and malaria. In studying the alarming inequities around the globe in the prevention and treatment of these diseases, Rozett argues that it is imperative that they be understood not only as a global health crisis, but also as a key source of political instability, economic depravation, social turmoil, and hence of threats to international peace.

In his consideration of the "health reform" movement of the past twenty-five years, Thomas Rathwell shows how the waves of reform in the health sector due to rising costs, poor quality of care, and consumer dissatisfaction around the world have in many cases been driven by a shift in thinking about the role of government in organizing and managing health care delivery. This shift towards "new public management" has been founded on the belief that a shift toward market-based competition in health care would lead to cost savings and better service. Rathwell analyzes this movement with a special focus on the issues of accountability, equity, and solidarity; he finds surprising variations in the degree to which these reforms have been effective within various national contexts.

Chinyelu K. Lee and Dorothy Brewster-Lee examine a similar set of issues in reflecting on the sustainable health care initiative—"Health For All by the Year 2000"—that emerged from the WHO's 1978 Alma Ata Conference on Primary Health Care. They use three case studies drawn from their experience in sub-Saharan Africa as a means of examining the development of sustainable, locally controlled, health care. In their view, the story of the work of the WHO and many of the larger non-governmental organizations (henceforth NGOs) over last two decades has been one of missed opportunities and misdirected efforts. They argue that locally rooted faith-based organizations will be a key tool in addressing the current and future crises in health care in Africa and around the world.

Socrates Litsios contribution to this section of the book is based on the comments that he gave at the conference on the current state of efforts to control malaria around the world. The failure of malaria control efforts, he argues, has been the by-product of a number of unfortunate decisions and misdirected efforts. He sees a number of critical problems that have developed in malaria control efforts over the past forty years, and contends that the only hope for progress in this area lies in investing in local communities while at the same time maintaining a vigorous international infrastructure that will be able to support local efforts to combat malaria.

Finally, Steven Latham takes the issue of accountability as fundamental in his philosophical reflections on the issue of global health. His argument relies on the idea that when the need is great enough, individuals come under a moral obligation to provide assistance to their fellow human beings. Extrapolating from this premise, Latham argues, a similar burden of responsibility can also be applied to corporate entities and private organizations. In particular, he argues, foundations involved in international health must be grounded in accountability to the people who are affected by their policies and decisions.

In his plenary remarks on "Racing Against the Global Industrial Hangover," Peter C. Goldmark, Jr. paints a disturbing picture of what he metaphorically describes as the coming storm. He argues that the past fifty years have been marked by an era of great wealth accumulation and technological development. On the horizon, however, loom major storm clouds in the form of a declining international order, the growing threat of weapons of mass destruction, and the increasing degradation of the global environment. These conditions, he asserts, have produced a dangerously unstable situation. But while the current climate is dire, we are not without options. For Goldmark, our best hope comes in the form of working to put into place a global basic health package that would provide a minimum level of health security to every individual on the planet. Such a program would be expensive—Goldmark estimates that it would cost 100 billion dollars a year—and foundations and other NGOs would need to play a leading role in its development and implementation. It would, however, provide hope to billions of people around the world. And hope, he argues, may be all that we have in the face of a desperately critical situation—it could even point us toward a better, more secure and more equitable world for all.

The papers in this collection, clearly, are diverse in their approach, in their specific concerns, and in their conclusions about the past and future role of philanthropic foundations in global healthcare. The discussions at the conference were lively, and the exchanges, although collegial, were sometimes heated. Several major themes emerged at the conference that we hope shine through the diversity of these papers. One such theme is clearly that of missed opportunities and misguided priorities. Although we are all aware of how messy situations can become much clearer when examined in hindsight, in the cases presented at the conference this theme was all too frequently evident. Another key theme that emerged in many papers at the conference was the perception that we are at a key inflection point in many global health issues. Several critical issues in the global health situation will unfold in the first three decades of the new millennium. We hope that the proceedings of this conference can be of use to those who will be making critical decisions about the future at foundations, public health agencies, government agencies and NGOs. It is our hope that in publishing these papers, others can capture something of the spirit of the conference and use whatever portions they may find valuable in their work. In the spirit of the Chinese poet Lu Xun quoted in Peter Goldmark's remarks, we see these works as efforts to lay down a path in the countryside, and we hope that others may follow.

Part One

Historical Considerations

Chapter One

Evaluation and Accountability: with a case study of the Early Rockefeller Foundation

Benjamin B. Page

The profit-oriented corporation and the philanthropic foundation—both have played fundamental roles in shaping our national and global communities. Yet each has a place within these communities that is rather anomalous. While operating in a world most of which is committed to democracy and its expansion, both corporations and foundations are controlled by small boards that are, in the one case, elected by shareholders from a list, usually nominated by a committee of the existing board or, in the other case, self-perpetuating. Chief Officers are in turn selected by and responsible to such boards; neither government or society at large, nor communities affected by the actions of corporations or foundations have access to or roles within their decision-making.

To be sure, corporations operate within parameters set by various regulations and laws, and have been subjected to periodic waves of public scrutiny, many of which their own actions have occasioned. Moreover, they are ultimately subjected to the iron hand of the bottom line: there is an ultimate, measurable purpose they are intended to serve and by which their operations can be evaluated. Foundations, on the other hand, have by and large been free from such limitations. The situation of foundations has been described as one in which, apart from the broad general direction set by their mission statements, and understandings regarding principles of governance and the use of financial resources, nonprofit organizations are "largely free to pursue their missions as their staff and boards please." There is no public agency with the authority to scrutinize their work, nor are there voters or shareholders whose opinions or questions need to be considered. As a nonprofit comes to depend more on income from endowments than on contributions, even large donors "recede as independent judges," while beneficiaries of programs are "distant and unempowered."[1]

Under such conditions any sort of objective evaluation of mission, program, performance, process, or outcomes becomes problematic, to say the least. This, turn, becomes an important issue for several reasons. In contrast to government and, increasingly, the corporate sector and the media, foundations enjoy a generally accepted reputation for doing good and a resulting rather high level of public trust. Thus they have been permitted to keep under their individual and collective control vast amounts of wealth, enhanced by tax benefits for both themselves and their donors, and to choose on their own initiative, again with very little by way of accountability to any outside body, the substantive and geographic areas in which to become involved. Given the considerable impact foundations can have—seek to have—in numerous areas of socio-economic well-being or in legislation or political agenda, the society in which foundations function has a legitimate interest in asking, at the very least, whether it is getting sufficient benefit in return—value for trust and tax privilege.

This is precisely what two significant publications from within the philanthropy community have found to be problematical. In *Evaluation for Foundations* (1993) the Council on Foundations began from the observation that while "evaluation of product and process in grantmaking programs" had long been seen as "especially worthy of encouragement," there were very few foundations or corporate giving programs that undertook "serious or sustained evaluation" of their own work or that of grantees, and that there were hardly any in the grant-making profession "sufficiently comfortable in their knowledge of evaluative practices as to be able to offer training or consultation to others". It thus offered its 1993 book of "concepts, cases, guidelines and resources" as a contribution to the process of enabling grant-makers to take "confident, successful *first steps* into program or process evaluation;" [2] the seventh of the thirty five "Keys To Effective Evaluating" warns potential evaluators not to be "overly impressed with what has been done or written so far in the evaluation profession." The editors of the conference volume of the American Assembly conference on *Philanthropy and the Nonprofit Sector in a Changing America* (1998) "reluctantly concluded, over the course of the Assembly conference, that too many of us in philanthropy and the nonprofit sector believe that simply because we know we are doing good the public should trust us." In contrast, they insist, "The public is entitled to continued evidence that its trust is warranted", and that "public accountability for philanthropy and the nonprofit sector should extend well beyond issues of taxability." In this vein, the conference's "Final Report" urged organizations in philanthropy and the nonprofit sector to "strive continually to improve their operating effectiveness and their accountability to the public." To "promote public trust and to motivate giving and volunteering" they need to "learn evaluation methods, conduct active evaluation programs, and communicate [including to the media and via the internet] the results" including their activities and accomplishments, even their tax forms. "They must work to earn and keep the trust of those they serve." [3] Otherwise, warns another author cited in volume, the current dearth of regulatory attention paid to accountability in the nonprofit sector may become "the prologue for major legislative attention—

attention that should be paid by those within philanthropy before others determine from outside what should be the case."[4]

The story of the various entities created out of the fortune of J.D. Rockefeller, although it began nearly a century before the above was written, is an example of the kind of situation decried in the recent discussion of evaluation and accountability—and an indication that that situation is far from new. To begin with, no individual or agency asked Rockefeller or his chief financial advisor, Frederick Gates—"the guiding genius behind all our giving"[5]—to create the Rockefeller Foundation (RF, 1913) or any of its precursors. Nor did anyone ask them to undertake what became their primary commitments for over two decades: the global spread of "civilization"—the culture, religion, attitudes toward work and money, and, above all, the spirit of "science" that they believed had had emerged from European experience as a universal norm—or, as a vehicle, the world-wide standardization of the "scientific model" of medicine, public health, and nursing that had been implied in Abraham Flexner's 1910 report on medical education.[6] The first step in these directions, the Rockefeller Institute for Medical Research (RIMR, 1901), emerged from Gates' reading of a book well outside his own background[7]—Osler's *Principles and Practice of Medicine*—and his enthusiasm for creating in the U.S. an institute like those in Europe associated with the work of Pasteur or Koch.[8] Although the Rockefellers had, in 1903, funded an agency to stimulate, basic education and teacher training in the US south, and—with the aim of building local tax bases to support local education—agricultural education and development (the General Education Board, GEB), Gates eight years later added quite a different undertaking to its mission: a campaign to use the promise of Rockefeller contributions to buildings or endowments as a lever to entice elite northern medical schools, despite often strenuous opposition from the profession, to require a full time commitment from their clinical faculty members[9]. In 1909, following a demonstration that the laziness endemic in the rural south was in fact due to an easily identifiable, treatable, and preventable disease, Gates convinced Rockefeller to create a Sanitary Commission for the Eradication of Hookworm (RSC). The RSC's first director, Wickliffe Rose promptly announced that he would not stop with hookworm, that efforts to counter it would serve as "an entering wedge": the hookworm campaign would lead ultimately to "consideration of the whole question of medical education, the organizing of systems of public health, and the training of men for the public health service."[10] In pursuit of this Rose together with Flexner, who was by then directing the GEB's "full time campaign,"[11] created at Johns Hopkins University the first graduate level school and curriculum for full time health officers (1916).[12]

As early as 1905 Gates had begun urging Rockefeller that, since his fortune "had been drawn more widely from mankind than [that] of any man who has preceded you, [a portion of it might be devoted to] the needs of all nations and the islands of the sea [and returned to them] laden . . . with the blessings of civilization and Christianity;"[13] Gates was soon to replace "Christianity" with "Science" and "scientific" medicine as the content of the missionary work into which he guided Rockefeller. Such thinking, plus a second 1905 letter from Gates—this one expressing to Rockefeller concern over the future of his fortune[14]—together with the finding that hookworm was a serious and preventable problem in a "belt" of countries that circled the globe and the growing temptation of China as a field ripe for the expansion of Western ideas, resulted in the idea of a great Foundation to, in the modest words of the charter Gates authored (1913), "promote the well-being of mankind throughout the world."[15]

The trustees of the newly-launched RF—most of them selected or approved by Gates—decided not to publish policies in order to preserve flexibility.[16] Thus for the first several years they groped to determine just what the charter required of—and permitted to—the entity over which they had charge. Their first act (June, 1913) was to create an International Health Board (IHB) with Rose as Director, to carry the hookworm and public health campaigns world wide, and the China Medical Board (CMB), to carry "scientific medicine" and medical education to that country.[17] Other areas were considered, from industrial relations and social hygiene to the fine arts. However a strike that turned bloody in a Colorado mine in which the Rockefellers had interests and an ensuing Congressional investigation led the trustees to conclude that potentially controversial areas were to be avoided,[18] while the outbreak of the First World War involved the RF in substantial efforts to aid civilian populations in German-occupied countries.[19] Until the end of 1919, then, except for the addition of a Commission for the Prevention of Tuberculosis in France (1917), the IHB and the CMB remained the only active undertakings of the RF. In later years Gates indicated that his only regret was that the charter had not restricted the RF to the areas of national and international health and medicine. Its work in these areas, he wrote, had met with nearly universal approval; when it entered other areas its involvement had been "widely and in some particulars not unfairly condemned."[20]

In December of 1917 George Vincent, newly installed as the second RF president, sketched a vision for the development of medical education "on a world-wide scale." Although once again there were no requests for anything of the sort, nor any studies of demonstrated need, a Division of Medical Education was created to implement the project (DME, 1919). The charge given the new entity was to focus initially on-going work in the US, China, and Brazil, and explore possibilities in Canada, Peru, Turkey, Australia, Thailand, India, Hong Kong, and South Africa. With medical education now on its own Rose was free to concentrate the work of the IHB on public health, foreseeing, in 1919, "a series of schools of hygiene at strategic sites all over the world, with relations of interchange and including the building up of public health organization, statistics and public health laboratory services, and demonstrations in the control of [various diseases]."[21] Although not mentioned in these mandates and visions, several countries of Europe soon came to be included in the purview of both agencies.[22]

With regard to the accomplishment of the visions put forth just after World War One, the Rockefeller projects can be termed successful. By 1950, through its IHB, CMB, and DME, the RF had planted institutions and training facilities propagating "scientific" medicine and public health, and exchanges for personnel from schools and health ministries and agencies, in the US, China, Canada and some ninety other countries in Central and South America, the Caribbean, the Near East, Asia, the Pacific, and Europe, and had made significant contributions to the development of the League of Nations Health Organization, UNRRA, and the UN's WHO.[23] Additionally, in the US, although the campaign had been terminated in 1925, the GEB had gone far toward establishing the "full time plan" as the norm in medical education, while Flexner's report with its stringent application of "the scientific model" to the teaching of medicine, had led to the closing of nearly half of U.S. medical schools and, thanks significantly to the promise of Rockefeller largesse, the conversion of the remainder to the education in the "scientific medicine" that the report had urged. The project of establishing globally the "scientific model" had also successfully accomplished a further, rarely-spoken purpose. As already noted, there was a second significant 1905 letter from Gates to Rockefeller. The first had begun moving Rockefeller toward international giving; the second had raised the prospect of a need to create a productive way to both use and perpetuate Rockefeller's fortune, lest it "crush you and your children and your children's children."[24] For this the various Rockefeller undertakings proved at least equally successful.

These very successes, however, point toward questions, including whether "success" in the realization of a foundation's usually self-appointed project implies that the project was "beneficial "or "effective", that it was the most useful way the financial and human resources involved could have been used, or that the foundation's mission itself was needed or at least useful to society at large. With respect to the RF, has the world-wide establishment of schools and institutions of "scientific" medicine, public health and nursing in fact contributed to the *health* of mankind. To be sure, this was never a primary goal; as already indicated, for Gates and the Rockefellers the ultimate goals were global spread of "civilization" and the spirit of "science," and the useful perpetuation of the Rockefeller fortune.[25] Like most ventures of Rockefeller agencies, the project in "scientific" medicine and public health was undertaken on the basis of assumptions, visions, and convictions of personnel and trustees of the RF itself, with little reference to requests for assistance, demonstrated need, or issues of evaluation. The case of the campaign for the "full time plan" seems to have been typical. Although the plan remained a *sine qua non* of RF support for schools and ministries world wide until 1925, a 1919 report indicated that it had been undertaken, in 1911, "on the presumption based on a priori consideration that the system is worth the price, [even though] it still remains to be objectively proven that teaching, the care of the sick, and scientific production are all so much better off under the full time scheme that universities generally should move to its adoption."[26] Nor did RF personnel often evaluate their efforts, once under way. Letters from field staff tend to attribute lack of success, local criticism, or opposition to personality clashes, vested interests, or politics. The account presented to a special meeting of RF trustees in 1933 is characteristic: in an early instance of what has come to be known as "blaming the victim," it declares "failures and disappointments" to be "most commonly associated with incompetent recipients, premature hopes on our part of progress, and over-confidence in the social, economic, and cultural matrix in certain countries."[27]

From another perspective, there have been numerous studies indicating that "scientific" medicine has in fact played at best a subsidiary role in the last two centuries' reductions of morbidity and mortality; far more has been due to general economic development and ensuing improvement in water supply, nutrition, housing, sanitation, education, and opportunity. Recent studies indicate a far greater role of non-biomedical factors even in today's main "agenda-setting diseases", tuberculosis, AIDS and malaria.[28] Along with studies calling for far more aid to economic development, there are others suggesting that the best way to enhance the general health status of a country may be direct economic assistance to individuals—improved pay or pensions for example, or micro-loans.[29] Moreover, the economic and human costs of the "scientific model of medicine" and its campaigns against disease based on laboratory research and resulting pharmaceuticals, technologies, and hospital-based care, all often with severe side effects, are being found overly burdensome even for economically developed countries. Hence the growing interest—on the part of "consumers" if not of institutions—in popular or traditional alternatives, including many the work of the RF had pushed into the background.[30]

While the "scientific model" has proven efficacious in numerous ways, much of this efficacy has been due to the large amounts of support given from the RF and other such agencies during the first half of the last century. To a large extent it can be said that it was the RF, more than any single entity that made possible the establishment of "scientific" medicine as "scientific" and, for much of the past 100 years, as the solely valid form of response to disease. So committed to the model were the men of the RF that they ignored, denigrated, or rode rough-shod over pre-existing elements or traditions of medical or public health theory, education, organization, or practice. A letter, hand-marked "Personal", from S.M. Gunn to RF president Vincent at the time Gunn was made a vice president (1927), while telling, is exceptional. Reflecting on his decade of RF experience in Europe, Gunn wondered "Are we not dogmatists and too much sold to our dogmas? Are we not developing the same weaknesses (?) as are attributed to missionaries? Too much faith in our own particular brands of public health, medical education, nursing, etc.

. . . [with a tendency] to rather pooh-pooh the points of view of many leading [experts] of other nationalities [including with respect to work in their own countries]."[31] Clearly there is room to wonder what kinds of contribution to individual, national, or global *health* we might see today, had a portion of the funds bestowed on "scientific" medicine, been devoted any of the extant alternatives—any of those Gunn mentioned or the work of the RF pushed aside—homeopathy, for example (which Rockefeller himself followed until his death at age ninety-eight) or the Chinese, Indian and other traditional medicines today re-emerging.

All of this points toward a related question: has the successful nearly global planting of the schools and institutions of "scientific" medicine, public health, and nursing been the most effective, beneficial way the RF could have pursued its self-given mission of promoting human well-being throughout the world, the most beneficial way the millions of dollars devoted to the project could have been spent. Both Rockefeller and the RF proclaimed interest in searching for root causes rather than responding to symptoms.[32] But *is* disease, as Gates insisted "the supreme ill of human life . . . the main source of almost all other human ills—crime, poverty, ignorance, vice, inefficiency, hereditary taint, and many other evils?"[33] In more recent years there has emerged a strong body of opinion to the effect that the root cause of the items on Gates's list or of such conditions as reduced life expectancy, malnutrition, handicap and disability, or the inability of individuals or communities to comply with medical recommendations or public health regulations is in fact poverty.[34] Nor was this view of the relationship between poverty and disease unknown even in the day of Gates and the shaping if the early RF, as evidenced by, for example the settlement house movement or the then increasingly vocal critique raised by the various movements for political reform and social change.[35] Such perspectives, however, were hardly mentioned within RF circles:[36] pursuing it could ultimately have opened questions about the health consequences of the economic system that made possible fortunes like Rockefeller's, and thus conflicted with one of Gates' objectives in the creation of the RF in the first place. All of this a focus on the biological etiology associated with the scientific model and its germ theory avoided. As with the possible public or individual health benefits that might have flowed from RF support of alternatives to "the scientific model", we can today do no more than raise the question of the possible benefits for human well-being that might have flowed from support for research into—and action against—economic, social, or cultural factors in health and disease, or from consideration of other possible ways of contributing to global human well-being. They are, nonetheless, issues that would need to be explored in any overall evaluation of the work of the RF.

Finally, in addition to such questions about the mission of a foundation and its pursuit thereof, there are questions about the place of a nonprofit in the larger society. This is particularly so in the case of foundations based on contributions or income from private fortunes. Who has benefited most from the privileged position of nonprofits, their large degree of freedom from external scrutiny, or their role in US and, increasingly, global society? Has the perpetuation of the large private fortunes—an outcome if not always explicitly a goal in the founding of foundations—been beneficial in the context of societies committed to democracy? To what extent does "democracy" include economic or socio-cultural factors as well as political arrangements? Ultimately, of course, such questions raise the issue of the implications of an economic system that permits the amassing of fortunes of the sort foundations are made of. Such questions, to be sure, go beyond the RF and are hardly new; they have been raised from time to time in the labor movement, the press, Congress—in the case of the RF even in the White House.[37] Although recently dormant, they are becoming increasingly pertinent given the increasing gap between the super rich and the rest of the world, the growth of tax and other benefits associated with wealth, threats to medical care and other social benefits for the less than wealthy, and an economic system that, in its inherent need to produce "growth" and profits increasingly risks the ecological future and militarization of the planet. Issues such as these raise serious problems for health and other aspects of "the wellbeing of mankind throughout the world"; they also seem to be inherent in the current, predominantly economic form of globalization. They are thus questions that need to be part of any evaluation of the work and accountability of foundations and other nonprofit organizations operating in the context of a world ostensibly committed to democracy and its spread.

Notes

1. Warren F. Ilchman and Dwight F. Burlingame, "Accountability in a Changing Philanthropic Environment," in *Philanthropy and the Nonprofit Sector in a Changing America*, ed. Charles T. Clotfelter and Thomas Ehrlich (Bloomington: Indiana University Press, 1999).

2. Humphrey Doermann in Council on Foundations, eds., *Evaluation for Foundations: Concepts, Cases, Guidelines, Resources*" (San Francisco: Jossey-Bass, 1993), xv (emphasis added).

3. Charles T. Clotfelter and Thomas Ehrlich, "The World We Must Build," in Clofelter and Ehrlich, *Philanthropy and the Nonprofit Sector*, 514; "Final Report", *ibid.*, 525-527.

4. Harvey Dale, Norman A. Sugarman Memorial Lecture, Mandel Center for Non-profit Organizations, Caser Western Reserve University, 1994, cited in Ilchman and Burlingame, *Philanthropy and the Nonprofit Sector*, 205.

5. Observation of Rockefeller, ca 1917, cited in Raymond B. Fosdick, *The Story of the Rockefeller Foundation,* (New York: Harpers, 1952), 7.

6. At the tenth anniversary of the RIMR (1911) Gates had declared that since all people share the desire for health, the values of medical research, were the most universal values on earth; because of its universal acceptance, scientific medicine could be used "to colonize and convert." Its 1920 Annual Report declared, the post-war program of the RF would be to "the

education of countries and nations . . . [seeking] to change habits of thought . . . and engender a new attitude toward disease and toward medical care." Gates, in RAC Gates Collection, Speech at Tenth Anniversary of the RIMR; RF *Annual Report*, 1920, 14.

7. Gates had been a Baptist clergyman and fund-raiser before moving to Rockefeller's office.

8. Fosdick, *The Story of the Rockefeller Foundation*, 8.

9. In 1911 Gates sent Flexner to Johns Hopkins to ask Dr. William Welch of the Johns Hopkins Medical School what he would do with a gift of one million dollars. The "full time plan" was built on Welch's reply.

10. Rose to RF secretary Jerome Greene, September 30, 1915, cited in the multi-volume documentary of itself prepared by the RF (hereafter RF History), 1278, Rockefeller Archive Center (RAC) 900-1; John Ettling, *The Germ of Laziness: Rockefeller Philanthropy and Public Health in the New South* (Cambridge, Harvard University Press, 1981); also Fosdick, *The Story of the Rockefeller Foundation*, 10.

11. Flexner directed the campaign for "the full time plan" from its inception, ca. 1911 until it was terminated, in 1927. Ironically, the plan was not among the criteria by which he evaluated medical schools when preparing his 1910 report on medical education.

12. On Rose's fundamental role in designing the school and curriculum, see Elizabeth Fee and Roy Acheson, eds., *A History of Education in Public Health* (Oxford University Press, 1991), 7.

13. Gates to Rockefeller, "Letter on Missions", 1905, cited in the multi-volume documentary history of itself prepared by the RF ca. 1938, Rockefeller Archive Center (RAC) RF900-901 *RF History*, 1249.

14. Gates to Rockefeller, June 3, 1905, in RAC Gates Collection, Papers.

15. Gates was a member of the board of the new Foundation--as he was of all initiatives with which he was associated--but never an officer. Its first President, JDR Jr moved to the Chair of the Board, a new office, and was succeeded by George Vincent, in 1917. Nonetheless, in the words of a future President--a board member from the outset, it was Gates who "set the tone for the Foundation in terms of human welfare in a world without boundaries." Fosdick, *The Story of the Rockefeller Foundation*, 20.

16. *RF History*, 1751.

17. Both Boards were initially called "Commissions".

18. *RF History*, 1475-81; Fosdick quotes Gates to this effect, but his interpretation was that it was direct RF involvement in potentially controversial areas that was to be avoided; in such cases, the RF would operate by means of funding other organizations that were already involved, the better to preserve objectivity. Fosdick, *The Story of the Rockefeller Foundation*, 20-29.

19. In 1917 RF president George Vincent observed that while war relief work was probably necessary and worthwhile, he felt it a detour from what the RF could have been doing "in a world of reason and sanity". Fosdick, *The Story of the Rockefeller Foundation.*, p. 28. Gates' observation, ca. 1923, is in RAC Gates Collection, Papers.

20. Gates, papers, memo written in 1923, cited in Fosdick, *The Story of the Rockefeller Foundation*, 28.

21. Vincent, "Outline of Possible Program Developments", December, 1917, cited in

RF History, 10: 2659; Rose, "Notes on a Journey to France, 19 March-10 May, 1919," entry for 19 April, p.9; RAC RF WR, 1916-1922.

22. Further, Paul Weindling, "Public Health and Political Stability: The Rockefeller Foundation in Central and Eastern Europe between the Two World Wars," *Minerva* 31 (1993), and Benjamin B. Page, "The Rockefeller Foundation and Central Europe: A Reconsideration", *Minerva* 40 (2002): 265-287.

23. Fosdick, *The Story of the Rockefeller Foundation*, 266, 314. These figures include the work and fellowships associated with the International Education Board, established in 1923, but intended to exist for only 5 years, by which time both principal and income were to have been exhausted. The CMB became independent of the RF in 1933; the figures given include the work or that agency only up until then.

24. Gates, note 13 above.

25. It has occasionally been suggested that there is a remarkable parallelism between areas of RF involvement and areas of Rockefeller Oil interest, but studies have so far found no "smoking gun." Cf. e.g. Susan Gross Solomon and Nikolai Krementsov, "Giving and Taking Across Borders: The Rockefeller Foundation and Russia, 1918-28," *Minerva* 39.3 (2001): 265-298.

26. *Annual Report of the General Education Board, 1918-1919,* cited in Fosdick, *Story of the Rockefeller Foundation*, 98.

27. From the Agenda prepared for a special meeting of RF trustees, April 11, 1933, p. 22, RAC RF 3.1/900/22/168. Cf. also Paul Farmer, *Pathologies of Power: Health, Human Rights, and the New War on the Poor* (Berkeley: University of California Press, 2003), 140, 147.

28. E.g. Thomas McKeown, *The Role of Medicine: Dream, Mirage, Nemesis* (Princeton University Press, 1979); Ronald Rozett, "Agenda-Setting Diseases," in the present collection.

29. E.g. Anne Case, "Health, income, and Economic Development", presented at the Annual Bank Conference on Development Economics, World bank, Washington, D.C., 2 May, 2001.

30. Cf e/.g. Warren Salmon, ed., Alternative Medicines: Popular and Policy Perspectives (New York: Tavistock, 1984).

31. Gunn to Vincent, 16 May 1927, RAC RF RG 3/900/21/59.1

32. John D. Rockefeller, *Random Reminiscences of Men and Event* (Garden City: Doubleday, 1937), 177.

33. RAC Gates Collection, Personal memo "Philanthropy and Civilization."

34. WHO *World Health Report*, Geneva, 1955 and subsequent years; Jim Yong Kim, et al., *Dying for Growth: Global Inequality and the Health of the Poor* (Monroe Maine, Common Courage Press, 2000), 14; Farmer, *Pathologies of Power.*

35. CF e.g. G.B. Shaw, "Preface on Doctors", in *Major Barbara* (various editions). In the US see e.g. the socialist movement of people like Eugene Debs or the settlement house movement associated with Jane Addams.

36. Socrates Litsios has described the outcome of the efforts of the person within the RF most strongly in favor of a more social approach to disease, in "Selsklar Gunn and Paul Russell of the Rockefeller Foundation: A Contrast in Styles," in the present collec-

tion.

37. Cf e.g. Congressional and press discussions over the unsuccessful effort of Rockefeller interests to obtain a charter for the RF from Congress (1910-13). President Taft's Attorney General, William Wickersham considered the proposal an unprecedented "indefinite scheme for perpetuating vast wealth… entirely inconsistent with the public interest"; Taft agreed, calling the proposal "an act to incorporate John D. Rockefeller." Harry F. Pringle, *Life and Times of William Howard Taft* (New York: Farrar and Rinehart, 1937), 2:622-3. The RF was chartered in the State of New York (1913).

Chapter Two

The Rockefeller Foundation's 1925 *Malaria* Film: A Case Study in Early Public Health Filmmaking

Marianne Fedunkiw

Films have been used as a tool in public health for more than eighty years.[1] The Rockefeller Foundation acknowledged film as an effective medium and funded the production of a number of films, as early as the 1920s. Perhaps best-known of the early films produced by the Foundation is *Unhooking the Hookworm*.

This article examines the Rockefeller Foundation's next foray into public health filmmaking, the 1925 anti-malaria film, *Malaria*. This case study outlines how and why the Foundation got involved in *Malaria*, the geographical and cultural scope of the film, its impact and its 'lifespan' as a public health tool.

Films recognized as a powerful weapon against disease

In a book published in 1927, authors W. Allen Daley and Hester Viney described films and cinematography as "two powerful weapons in the hands of doctors and nurses in their agelong [sic] warfare against disease."[2] They cited the ability of film to explain complex ideas without requiring a wealth of background knowledge as well as film's ability to show both cause and effect "within the space of a pleasantly occupied half hour."[3]

Public health officials capitalised on the popularity of film to get their message across to the greatest number of people at a single sitting. People who would not readily attend a public health lecture would throng to see a film—Timothy Boon cites the example of more than 2,500 attending the screening of a venereal disease film in Blackburn, a town northwest of Manchester, England.[4] Even more telling is the example of a sexual hygiene film, *The End of the Road*, made in 1918. This silent feature warned of the dangers of promiscuity, venereal disease and pregnancy "out of wedlock" and was designed for young women who lived near American World War I training camps or who did war work.

When this old silent film was shown over a period of ten days in Toronto in 1932, almost 20,000 people paid admission to see it.[5] Films were restricted only by venue and the need for working projection equipment. Adolf Nichtenhauser offers one description of how so many saw these films:

> Health agencies and schools exhibited these films in school auditoriums, public halls, churches, rented movie houses, and sometimes in parks or streets on summer evenings. Sometimes, they were also included in the regular shows of the film theaters which then were more varied and independent in their programming than today.[6]

Boon estimates that, in Britain, 115 health films were made between 1911 and 1930[7] and more than double that number in the 1930s alone. The most popular topics of public health films before 1930 were hygiene, particularly sexual hygiene, operative techniques,[8] and venereal disease. Malaria films made up a small percentage of the total films produced.[9] In fact, of the films which I have found to date,[10] I have found only two malaria prevention 'features' that pre-date the Rockefeller Foundation-sponsored film.[11] This finding is corroborated by Nichtenhauser who wrote that, in the U.S., although "by the end of 1925, the number of health films had increased to over 300,"[12] there were only eight films dealing with the malaria mosquito compared to fifty-two on milk production.[13]

Learning from *Unhooking the Hookworm*

The Rockefeller Foundation made its first foray into feature-length public health motion pictures with *Unhooking the Hookworm*. The International Health Board authorised the making of this motion picture on 24 March 1920 and allocated a total of $4,800 for the single reel feature.[14] In an internal Foundation memorandum, the process was summarized as follows:

> After careful consideration the International Health Board decided several years ago to undertake the production of a hookworm film. Difficulties of various kinds were encountered but the work was carried on and the film was finally ready for use in the early part of 1921. Although not a popular subject of course treated in a popular way, the picture has attracted much attention. The results of its use in the southern states have been particularly satisfactory.[15]

The memo goes on to give a balanced overview of the advantages and disadvantages of film as a public health tool. The unidentified author of the memo addresses both the successful domestic use of the film as well as the more limited success in other countries. A state health officer, also unnamed, is glowing in his praise of the film:

> The hookworm film is far and away the best public film I have ever seen. You will be interested in knowing that as a result of showing it in the picture shows of—[sic] more than 500 specimens were submitted to our own laboratory, and that the physicians of— [sic] all report a large number of patients who have come to them for examination. Quite a considerable number have been treated

> and are being relieved. Your film is so excellent and has done such a vast amount of good I feel I cannot say too much in its praise.[16]

The scenes in the hookworm film were all distinctly American.[17] This did not, however, prevent the titles from being translated into Portuguese and Spanish (the Spanish title was *Luchando la Uncinaria*) although there were concerns about racial and cultural sensitivity. As W.C. Hausher[18] suggested,

> Races other than the white race are easily distracted when a partially exposed white person is pictured. It invariably brings down a roar of laughter and continually distracts the audience from the point that is being made. I do not believe in elimination of the white patients, because then many white folks will continue to keep the impression that hookworm does not affect the white people. Introduction of pictures of various sick patients from various countries would improve the film as nearly all races would then be represented.[19]

Others echoed Hausher and suggested: cutting or replacing "scenes and legends referring to 'Our Southern Cotton Fields;'"[20] for use in Ceylon, using "another game than 'going swimming', e.g. 1. Hackney racing 2. Mazoka;"[21] using the term 'privy' instead of 'latrine' because "[t]he people I have worked with don't know what a latrine is;"[22] and addressing the idea that, "[i]f the improved family conditions of a colored family were shown, it would leave a stronger impression than the white man's house in the closing picture."[23]

Copies of the film were loaned, free of charge, to individuals and health departments and other agencies within the United States to preview before purchasing a copy. By 1922, forty-four copies of the film had been sold, half within the U.S. and half "for use in foreign countries."[24] Two years later, total sales had more than doubled.[25] Sales were broken down as follows: English-language, domestic sales[26]—sixty-nine copies; English-language, foreign sales[27]—thirteen copies; Portuguese-language—eight copies, all to Rio de Janeiro, Brazil; and Spanish-language[28]—seven copies. The established price was $80.00, which represented the cost of making a positive print on non-inflammable stock.[29] Sales were handled directly by the IHB whereas rentals were handled by the Society for Visual Education.[30] By 1931, copies of the film were available in 35 mm format (either $85.00 or $35.00 on regular professional stock or $35.00 on safety stock) or in 16 mm format ($8.25).[31] The reason I give such detailed data on the sale of *Unhooking the Hookworm* is that I have not been able to find correspondingly detailed data on *Malaria*. I am sure, however, that sales of the malaria-prevention feature equalled if not surpassed that of the anti-hookworm film with parallels in geographic distribution. In much of the archival material, these films are discussed as a pair.

Getting *Malaria* made

The idea of the Rockefeller Foundation producing a public health feature on malaria dates back to at least October 1922. With the experience gained from making *Unhooking the Hookworm*, Foundation officers were trying to establish the need, scope, cost and style of a feature on malaria. Archival material offers a two-page summary of these early days wherein the anonymous author writes:

> Attention has been called from time to time to the need for a moving picture film for popular education on the nature, cause, and methods of control of malaria. No satisfactory film has yet been produced.[32]

Recommendations had been made regarding other proposed films[33] but the Foundation officers decided that the time was right to produce an up-to-date film that could be reproduced as required, and that could be used in various countries by simply translating the titles. The Foundation found that a former IHB employee, Edward Stuart, had submitted a more suitable scenario "which seemed to offer a basis for the kind of film which it might be advisable for the Board to produce."[34] After revising the script three times, the Board was ready to push forward toward production.

Toward that end, eight suggestions that covered the length, costs, languages, distribution and reproduction responsibilities, and staffing costs to produce a two-reel (2,000 feet) malaria feature were documented.[35] Having learned from the hookworm filmmaking experience, it was proposed to make this two-reel malaria feature using an outside agency "with only general supervision of the staff," for $5,000. This was only marginally more than the $4,402 spent on the single-reel hookworm feature. An additional $5,500 was earmarked to meet the travelling expenses related to field work for the malaria film.[36]

On 7 October 1922, Edward Stuart, Vice-President of Society for Visual Education, Inc. of New York City, sent a proposal to C.C. Williamson, Director of Information Service, that covered the cost of production, storage and distribution of both the original negative and positive copies, and distribution details.[37] Less than two weeks later, this contract was discussed by members of the Foundation. It was decided that although the main points were agreeable to the Foundation, "further negotiation seems to be necessary" on the cost of additional language sub-titles, the cost of prints, and the logistics and cost of distribution of both preview copies and sold copies.[38] Discussion ensued through until the end of 1922. In the end, it would be the Society for Visual Education that would make the film, a composite of new footage and parts of another film, *Malaria and the Mosquito*.[39]

Because of sensitivities to the latest medical research and techniques, cultural issues and distribution issues, and many script edits, *Malaria* took more than two years to complete. The script or scenario went through at least five edits. Copies were sent to various experts who were asked to read through the scenario and send on any comments, edits or suggestions. For example, Edwin O. Jordan was sent a copy of the draft in 1922. He compiled his suggestions in a letter, dated 8 September 1922, to Stuart.[40] Among Jordan's suggestions were giving a higher figure for the number of malaria cases in the U.S.—the draft title read, "Each year about 1,000,000 people in the United States suffer from malaria"[41]—and making use of a 1919 article in the *Journal of the American Medical Association* for "some interesting data on the cost of malaria."[42]

On 4 January 1923, copies of the third draft were sent to twelve physicians and professors,[43] two lay public health officials (U.S. Public Health Service in Washington, D.C. and the U.S. Bureau of Entomology in Mound, Louisiana), and Stuart.[44] An additional twenty-five copies were sent out to other physicians, Foundation staff and as extra mark-up copies.[45] The marked-up copies were returned between 16 January and 23 May 1923.

Some of the most interesting suggestions for revisions came from Alan Gregg,[46] at that time Associate Director, Division of Medical Education.[47] His first suggestion addressed the potential audience and potential impact of the film. Gregg suggested changing the sixth title, which read, "[number to be filled in] times as many persons are killed by malaria each year in India."[48] His reasoning was, "the reaction is 'Oh well what do I care as long as it was *India*'! '*The world*' would be better."[49] When the film was completed, the figure was filled in ("twenty times") and Gregg's suggestion of replacing "India" with "the world" was taken up.

Gregg was also sensitive to the potential audience and their apparent poverty and lack of understanding of modern science. First, he suggested that "ignorant people" might need a clearer explanation of the scene wherein a doctor examines a blood smear under the microscope:

> In less 'progessive' countries, few people know what a microscope does. I suggest a statement of its powers of augmentation with perhaps illust. of what a fine hair looks like at 800 diameters. Most ignorant people dont (sic) know difference between microscope and a spectroscope or X-ray.[50]

He also took exception to showing modern machinery being used to dig ditches that would drain marshes of standing water, the anopheles mosquito's preferred breeding site. In the scenario draft, the title (number 81) read: "SOME MARSHY PLACES MAY BE DRAINED IN A VARIETY OF WAYS" and the scene (number 60) was to show "Workmen digging several types of ditches. (Scene showing steam shovel in an elaborate piece of work). Scene showing a farmer digging small trenches in a field using Jersey ditch digger."[51] Gregg wrote, "[s]hould prefer to have *no* modern or complicated machinery connected with the idea of malarial drainage. The native or ignorant mind inevitably will infer novel machinery is necessary: it isn't."[52] In the film, the title read, "Marshy places and ponds may be drained in a variety of ways" and was followed by "scenes showing different types of ditches."[53] Gregg also suggested changes to aspects of preventive measures to be depicted in the film. In reference to the scene showing the proper use of protective netting on a bed, he noted: "Dr. S.T. Darling[54] insists that tucking under mattress brings netting too near arms and legs. Bar should go down all around bed and be down from early P.M on so that mosquitoes would hide under bed so enclosed."[55]

Of course, not all revisions were accepted. This last suggestion was disregarded[56] and in the final version of the film, the title read, "If the house is not screened mosquito bars give some protection, but they should be carefully tucked in." and the scene showed "Bed with mosquito bar. Man in pajamas carefully getting into bed, properly tucking the bar under the mattress."[57]

While the scenarios were being revised, others were concerned about getting the best possible images. On 16 April 1923, Williamson wrote to Stuart:

> Doctor Russell[58] also approved your idea in regard to additional malaria pictures. We understand it to be your proposal that while you are in the South working out the two reel malaria film, following the scenario which has been approved, you will take other things of special value that you may come across, showing more of the technical details of malaria control work, including drainage, fish control, and the use of larvicides, etc. While you are proposing to do this on your own responsibility, Doctor Russell thinks that if you should get enough material for an additional reel that would be of special use in the Board's malaria work, we might decide to purchase it.[59]

Another issue to be dealt with was the foreign language translations. The archives include correspondence addressing this aspect of *Malaria*. For example, in a letter to Professor H. C. Heaton in New York City, Williamson noted that, already in April 1923, the Foundation intended to issue the film in "four languages besides English, one of them being French."[60] He asked Heaton to proof read the French 'loop titles'. Williamson assured Heaton that the titles were "for the most part quite simple and non-technical" and went on to lament that, "[i]t happens that among the people to whom we turn for advice on medical matters there seems to be no one who is a native Frenchman."[61] The Spanish translation of the *Malaria* titles was completed within six months.[62]

Lifespan and impact of the film

The Rockefeller Foundation paid George E. Stone of Carmel, California[63] $3,722.00 for:

> All right, title and interest and all copyright in and to/that two reel motion picture film entitled "Malaria and the Mosquito" together with two copies thereof; the negative and the before mentioned two prints of which have already been delivered to the parties of the second part.[64]

Copyright was assigned to the Rockefeller Foundation on 16 April 1924 in Monterey, California.[65] The certificate was received by the Copyright Office of the Library of Congress in Washington, D.C. on 9 May 1924.[66] Credit for the scenario or script seems to have gone jointly to Darling, Professor William B. Herms,[67] Dr. E.H. Kleinschmidt[68] and Stuart.[69]

For those who were interested in purchasing the film, "Malaria: An Educational Motion Picture," a twelve-page booklet giving all the titles and descriptions of the images was produced by the IHB.[70] The film was divided into three 'parts.' The first part, used twenty-one titles and gave a general perspective on the geographical scope of malaria, symptoms, economic impact on farming, and the malaria parasite. "Part II," with thirty-eight titles, offered greater depth on the life cycle of the parasite while in the mosquito's gut and in the bloodstream of a human. It also showed the life cycle of the mosquito as an introduction to destroying the malaria vector. "Part III" (forty-two titles) dealt with mosquito eradication techniques. These included "spreading oil or poison" on stagnant ponds where mosquitoes breed, "stocking ponds with top minnows," "ditching by dynamite," clearing vegetation as well as using screens on windows and doors to minimise contact between man and mosquito and destroying the parasite with quinine.[71]

In addition to getting worldwide exposure, *Malaria* also encouraged other countries and organizations to use film as a form of public health propaganda. They not only used excerpts from the Foundation film but also adopted some of the practices used in the making of *Malaria*, including asking leading public health officials to vet scripts. One example was the Government of Porto (sic) Rico, Department of Health, Bureau of Malaria Control. They had produced a malaria film in 1931. In a letter to Dr. H.H. Howard of the Rockefeller Foundation,[72] Dr. Walter C. Earle wrote:

> I am enclosing a copy of the titles for the malaria picture we have been making here and I shall appreciate any criticism you may care to make. The pictures have all been taken and now we are waiting on Dr. Fernos for the Spanish translation. The day after we took the last picture, Viguié, the photographer, came down with malaria himself.[73]

In his reply, Howard told Earle that "these titles indicate to me that you have made a very good film . . . I would like to have the opportunity later of running the film through our machine here at the home office."[74]

There is also documentation of an offer to the State Health Officer in Alabama, as late as 1936, of "the scientific parts . . . in Reel II, . . . We should be perfectly willing to have you use any or all of this scientific part in any film that you may make locally and without any acknowledgement in the film" but stating that the film was "very badly out of date" and the Foundation "decided not to revise the film."[75]

Out of date

By mid-1931, some deemed both *Unhooking the Hookworm* and *Malaria* to be of limited value. As Dean explained in a letter to the American Medical Association:

> Both of these films were produced as aids in hookworm and malaria campaigns when the Foundation was carrying on campaigns in co-operation with foreign governments. The demand for these films is now relatively small and we consider that they have practically served the purpose for which they were pro-

> duced. We rather hesitate, therefore, to have them appear in any list that will be available to the layman.[76]

Dean was, however, willing to offer copies of the films for sale to physicians as late as August 1931. In fact, Dean wrote to the American Medical Association asking that both films be included in the AMA catalogue of motion picture films and also giving the "present prices" for positive prints on professional stock, Safety stock, and 16 mm formats.[77]

Perhaps the best summary of how best to use the still-pertinent footage of *Malaria*, is found in a letter from Dr. Mark F. Boyd of the Foundation's International Health Division to Foundation officer Dr. John A. Ferrell. Boyd criticised the film for being "for the most part considerably above the intellectual capacity of the average rural audience either in the United States or abroad," for being racially inappropriate for "rural negro audiences", and acknowledged the difficulty of making one film that could be used throughout the world.[78]

> No one realizes better than a southern negro the vast gulf that exists between the whites and the negroes, consequently the negro is not much impressed by scenes dealing with whites. On this assumption I duplicated the human interest scenario, with the cooperation of some negro actors, and cut out certain sections in one of the Foundation films and substituted these negro pictures. This modified film was enthusiastically received by negro audiences wherever it was shown and I am sure made the rural negroes appreciate malaria as a negro problem.[79]

This quotation is the best clue to why the Rockefeller Foundation did not want to be associated with the film less than ten years after it was completed: although malaria prevention techniques and the idea of the 'latest developments' in malaria prevention were cited, I do not believe this to be the true reason. In fact, the decade 1925-1935 was one of the least productive in terms of new malaria prevention techniques. There were no major advances. DDT and many of the newer pharmaceuticals would not come into play until the early 1940s. But problems involving race and cultural issues were making the film an issue with Foundation staff.

Boyd did, however, value recycling "the more technical portions, or the portions in which the photography is most difficult."[80] He advocated an arrangement whereby the Foundation would supply footage of these "technical portions" and a "human interest scenario" in 16 mm format to Foundation staff members throughout the world, asking the staff members to add local human interest material and footage of local malaria control operations.[81] Despite Boyd's criticisms of *Malaria*, he championed film as a public health tool. He ended his three-page letter with the comment that:

> I may say that it is my personal opinion that the movie is the most effective means of propaganda for reaching adults in isolated areas. This is made convenient and practicable by the use of a 16 mm film exhibited by a hand projec-

> tor with illumination supplied by a 6-volt automobile battery. The outfit is cheap, simple to operate, and can be transported into most remote regions where people have never before seen a movie.[82]

The following month, *Malaria* was screened for eight people.[83] These Foundation officers were trying to figure out not only what to do with the *Malaria* feature but also the idea of future films being produced by the Foundation. 'LTC' made the point that films are like textbooks and require constant revision to remain up-to-date. "If the RF policy is to put out films (sic) will have constantly to be reedited and brought up to date. Doubt whether call is great enough to continue. If it is let each man be responsible," 'LTC' noted.[84] Others continued to argue to salvage some of the material[85] but in the end,

> The consensus of opinion was that the technical part of the malaria film was satisfactory, but the other parts of the film were very poor and not in accordance with modern ideas of malaria control and that the entire film needs to be revised or its use discontinued.[86]

The films were withdrawn from general distribution although it was acknowledged that some segments still had currency. For example, in a bid to make the films available to the government as references and even as raw footage from which to glean salient material, W.A. Sawyer wrote to the Surgeon General in Washington, D.C. in 1937. He stated that both films "are now somewhat out of date and we have ceased to distribute them" but "our Scientific Directors will probably be inclined to present the negatives to the Public Health Service in case it desired to create new films in connection with its public health propaganda."[87] Sawyer went on to ask if there was any interest in this offer and, if not, if the Surgeon General might suggest "any solution for making a suitable use of the good parts of the negatives which were produced by the Division at considerable expense?"[88]

Malaria and *Unhooking the Hookworm* were donated to the U.S Public Health Service in 1937[89] and organizations seeking footage were directed to the federal office. Requests continued to come from around the globe, including India and Guam. In 1938, the Madras Public Health Department asked the Surgeon General for copies of the film, acknowledged as dated, for use in six Rural Sanitation Units in the Indian province.[90] Dr. W. P. Jacocks, the Foundation's Representative in India, was given permission to sell copies of the film to local Indian health organizations on the condition that no reference was made, in the film, to the International Health Division, and that the Division's name not appear anywhere in the film.[91]

Another request was made by J.T. Alexander, Governor of Guam. Alexander wrote asking for films or slides on hookworm and tuberculosis which could be shown on film trucks or in local cinemas.[92] In his reply, Sawyer explained that the hookworm and malaria films "became scientifically out of date" and the negatives had been given to the U.S. Public Health Service so that "any useful parts they might contain could be incorporated in future films prepared under the auspices of that service."[93] Sawyer went on to state:

> In deciding not to produce new films we were motivated in part by the diminished demand. It was becoming apparent that films showing local situations and people were more effective in foreign countries than those showing American scenes. The highly-developed program in Java is carried on with films and charts produced there.[94]

In this way, issues of race and culture could be dealt with by indigenous filmmakers, thereby exonerating the Foundation of potentially embarrassing sociocultural blunders.

The Rockefeller Foundation as public health propaganda filmmakers

In a memorandum dated 1936, G.H. Hazlehurst, Chief Engineer and Director, Bureau of Sanitation cited the value of motion pictures as public health tools. Hazlehurst explained that, in Alabama, motion pictures were used in public health education since 1921 and "with the development of 16 mm equipment, which was less expensive and easier to operate than the 35 mm type, the interest in the use of motion pictures was revived." Hazlehurst went on to say that more than half of fifty-six county health departments have 16 mm equipment[95] and that the malaria film was purchased by twenty-two countries.[96] Hazlehurst ended his memo with a plea for new films on hookworm and malaria: "It is believed that our County Health Units will continue the use of visual education provided suitable films are available, for if public health is to progress education must be kept up year after year."[97]

But it would take almost two decades before the Rockefeller Foundation would resume a role in malaria public filmmaking. Between 1943 and 1950, the Foundation helped produce five additional anti-malaria films. The Foundation joined with Walt Disney to produce *The Winged Scourge*.[98] Released in 1943, this co-production with the Co-ordinator of Inter-American Affairs, Nelson Rockefeller, showed the effects of malaria and used the Seven Dwarfs to illustrate how to get rid of potential malaria-carrying mosquitoes.[99]

This was followed by a series of Spanish-language animated Disney shorts, among which was *Insectos que transmitten enfermedes* (1945);[100] two black and white documentaries on the post-war Italian mosquito eradication project—*The Sardinian Project* (1949)[101] and *Adventure in Sardinia* (1950); and *War on Mosquitoes* (1950), a documentary on the mosquito eradication efforts in Ceara and Rio Grande do Norte, Brazil.[102]

When an inventory of film footage held in the International Health Division storeroom was done in November 1949, there were twenty reels on malaria, including three 35 mm sets of the 1925 film.[103] Officially, however, the Rockefeller Foundation had decided by the early 1950s that film making would be left to others. Despite the success of *Malaria* and the other malaria films in which it had some role, the Foundation could not be persuaded to undertake another malaria public health film in which it had such a dominant role. As H.B. van Wesep[104] wrote in a letter, dated 11 January 1952:

> The Foundation does not make a regular practice of producing documentary films. In the past, films have been made now and then of some of the Foundation's work, particularly that in hookworm and malaria, but these films became obsolete and are now not available.[105]

I believe that *Malaria* was the Rockefeller Foundation's last first-hand experiment with motion pictures as a tool in public health. The Foundation acknowledged film's potential for teaching and reaching large audiences but also noted the problems of interpretation and cultural scope when films were to be used on a global scale. The Foundation learned first-hand that language translations were not enough—images of indigenous peoples were desired, some argued were necessary, for use in individual countries. Although the official reason given for discontinuing distribution of the malaria film was that much of the material was out of date, I believe it was a combination of race and cultural issues that made the Foundation officers uneasy. There were no major developments in malaria prevention or prophylaxis in the decade following the film, but there is considerable discussion among the Foundation staff about the people and scenes shown in the film.

Unhooking the Hookworm was the only film to be funded and produced by Foundation staff. With *Malaria* the Foundation was already acknowledging that its officers were not filmmakers, but the Foundation's enthusiasm for film and anti-malarial initiatives was strong enough for the Foundation to be involved from the script-drafting stage to distribution. After *Malaria*, the Rockefeller Foundation would sponsor selected other films, but its involvement would be behind the scenes and would always involve local health or government organizations.

Notes

Excerpts from this paper were presented at "Batir de nouveaux ponts: sources, methods et interdisciplinarité/Building New Bridges: Sources, Methods, and Interdisciplinarity" held in Ottawa 8-9 May 2004 and at the American Association for the History of Medicine meeting held in Boston, 1-4 May 2003. This article was researched and written while on a postdoctoral fellowship funded by Associated Medical Services/Hannah Institute for the History of Medicine and held at the Wellcome Unit for the History of Medicine at Oxford University (WUHMO). The research was carried out at the Rockefeller Archive Center (RAC) supported by a 2002 RAC research grant. The author would like to thank: Michele Hiltzik, Carol Radovich and Dr Darwin H. Stapleton at the Rockefeller Archive Center; Dr Elizabeth Fee, Dr Steve Greenberg and Dr Michael Sappol at the National Library of Medicine; Prof Martin S. Pernick, University of Michigan; Dr Mark Harrison at the WUHMO, University of Oxford; and Prof Jacalyn Duffin for her continued support.

1. Among those who have published on public health films are: Timothy Martyn Boon, *Films and the Contestation of Public Health in Interwar Britain* (Ph.D. dissertation: University of London, 1999); Robert Eberwein, *Sex Ed: Film, Video, and the*

Framework of Desire (New Brunswick, N.J.: Rutgers, 1999); Marianne Fedunkiw, "Malaria Films: Motion Pictures as a Public Health Tool", *American Journal of Public Health* 93 (2003):1046-1056; Helen Harrison, (Ph.D. dissertation: Queen's University [Kingston, ON], 2002); Susan Lederer and John Parascandola, "Screening Syphilis: 'Dr. Ehrlich's Magic Bullet' Meets the Public Health Service," *Journal of the History of Medicine,* 53 (1998): 345-370; John Parascandola, "VD at the Movies: PHS Films of the 1930s and 1940s," *Public Health Reports 111* (1996):173-175; and Martin Pernick, *Black Stork: Eugenics and Death of 'Defective' Babies in American Medicine and Motion Pictures Since 1915* (New York: Oxford, 1996). There is also a 103-page report, "Film, and Young Peoples' Views on Health 1960s-1990s" by Virginia Berridge and Kelly Loughlin of the London School of Hygiene and Tropical Medicine, 1997 which is based on 48 films and videos taken from the larger *Catalogue of young peoples' views of health on film, 1946-1997.*

2. W. Allen Daley and Hester Viney, *Popular Education in Public Health* (London: H.K. Lewis, 1927), 109. At the time this book was published, Daley was a British Medical Officer of Health and Viney was a nurse.

3. Daley and Viney, *Popular Education* (n. 2), 110.

4. Boon, *Films and the Contestation of Public Health* (n. 1), 58. Daley and Viney expressed this in *Popular Education* (n. 2), 110. Films were "a very popular form of amusement, and a large audience who would not desire to hear a lecture will throng to see a film."

5. Adolf Nichtenhauser, *History of Motion Pictures in Medicine* (unpublished manuscript, circa 1953), National Library of Medicine, Adolf Nichtenhauser, Box 1—MS C 380, 1: 212. The film was shown in Toronto by Dr. Gordon Bates, General Director of the Canadian Social Hygiene Council. By 1926 alone, *The End of the Road* was reported to have been seen by more than eight million people.

6. Nichtenhauser, *History of Motion Pictures* (n. 5), 2: 70.

7. Boon, *Films and the Contestation of Public Health* (n. 1), 43-44. What Boon writes is that approximately 350 health films were made, in Britain, between 1911 and 1939 and that two-thirds of this total was produced in the 1930s. He also gives an estimated cost for an eight-minute film as £1,000 although low-budget operations could produce an eight-minute film for as little as £150.

8. In his section on medical films in the mid-1920s, Nichtenhauser states that "the great majority of all American films of the time still dealt with operative techniques," Nichtenhauser, *History of Motion Pictures* (n. 5), 2: 150.

9. I have found eleven films made between 1912 and 1930 which address malaria prevention or mosquito eradication. The anti-malaria 'feature' film output explodes with the beginning of World War II. Twenty-six anti-malaria films were produced between 1940 and 1945, the majority by the U.S. and U.K. armed forces for use in military education. This is an increase of more than forty percent and that is comparing a period of twenty-seven years with just five years.

10. I make this distinction because malaria preventive measures are shown in a number of pre-1925 newsreels, namely: *With the KUT Relief Force in Mesopotamia* (produced by Gaumont—1916); *World's Greatest Story* (War Office/New Era Film—1919); and *Pictorial History of the World War No. 10* (U.S. Signal Corps—1919). All three of these clips show men wearing mosquito hoods to avoid being bitten by malaria-carrying anopheles mosquitoes. I viewed copies of this footage at the Imperial War Museum in London, U.K.

11. In my research, I came across two of the earliest malaria films which focus on the mosquito: *The Mosquito (Science and Nature Series)* (Pathé—1912) and *The War on the Mosquito* (Edison—1912). Viewing copies of these films are available at the British Film Institute in London, U.K. Martin S. Pernick has found an additional eight films that deal with mosquitoes and malaria: *Extermination of Mosquitoes by Spraying Swamps* (Bray—1910s); *Sanitation in Mining Villages* (US Bureau of Mines/NJ Zinc Company—1916); *The Mosquito* (Army Medical Museum #147—1918); *Mosquito Eradication* (Army Medical Museum #154—1918); *The Mosquito Pest* (W.L. Brand—1919); *Mosquito Control* (1921—other details unknown); *Exterminate the Mosquito* (US Bureau of Mines—1921); and *How the Mosquito Spreads Disease* (a 17-minute film that was part of *The Science of Life* series produced by Bray—1922—see n. 13). Thanks to Professor Pernick for these additional films.

12. Nichtenhauser, *History of Motion Pictures* (n. 5), 3: 63.

13. *Ibid.*, 66. One of the early malaria films cited by Nichtenhauser is *How the Mosquito Spreads Disease,* produced by Bray for $18,000 in 1921 for the U.S. Public Health Service. This film was part of a series of twelve films that "sold well" in the Near East, South Africa and Japan and some of which were translated into Spanish for distribution in South America (2: 98-109).

14. Rockefeller Archive Center, RF, Record Group 1, Series 100, Box 5, Folder 43, bound volume, Memorandum, "Malaria Film", p. 2. The film was "a little less than 1000 feet in length" and was completed under budget: $4,402 of the allotted $4,800 was used to make the hookworm feature.

15. Rockefeller Archive Center, RF, Record Group 1.1, Series 100, Box 5, Folder 40, Memorandum, "Unhooking the Hookworm", p. 1. Although this memo is neither dated nor signed, I suspect that it is part of a longer report which is dated 26 May 1921 and has the initials "ECM:EH" under the date. This part of a memo, titled "Memorandum concerning Comments on the film *Unhooking the Hookworm*", is also part of Rockefeller Archive Center, RF, Record Group 1.1, Series 100, Box 5, Folder 40, and includes comments on the film, suggestions for improvement, including adding a cartoon segment because of the "tremendous sensation which the comic cartoon pictures make in the theatre in these lands" (p. 4). The comments included in this 1921 memo come from American state health officers, a professor of preventive medicine and hygiene at University of Georgia, a motion picture industry executive, a representative of the Russell Sage Foundation, and the chief of the Division of Popular Health Instruction, League of Red Cross Societies in Geneva, Switzerland.

16. Rockefeller Archive Center, RF, Record Group 1.1, Series 100, Box 5, Folder 40, Memorandum, "Unhooking the Hookworm" (n. 15), p. 1.

17. A full list of the titles in the film can be found in Rockefeller Archive Center, RF, Record Group 1.1, Series 100, Box 5, Folder 39, 20 April 1938, Memorandum from Major-General N.M. Wilson, Director of Public Health, Madras, to Dr. W.P Jacocks, Representative of the Rockefeller Foundation in India, pp. 2-3.

18. Walter C. Hausher (also spelled 'Hausheer') served with the Foundation's International Health Board, working on hookworm control in Mississippi, Trinidad and Tobago, Dutch Guiana and Jamaica, from 1919 to 1925. He resigned his position as Field Director on 31 December 1928. Thanks to Michele Hiltzik, archivist at the Rockefeller Archive Center, for this and other staff information throughout this paper.

19. Rockefeller Archive Center, RF, Record Group 1.1, Series 100, Box 5, Folder 40, Memorandum, "Suggestions for the Improvement of the Hookworm Film", p. 1.

20. Rockefeller Archive Center, RF, Record Group 1.1, Series 100, Box 5, Folder 40, Memorandum, "Suggestions for the Improvement of the Hookworm Film" (n. 19), p. 1. Suggestion of W.A. Sawyer regarding 'Subtitle 5'.

21. Rockefeller Archive Center, RF, Record Group 1.1, Series 100, Box 5, Folder 40, Memorandum, "Suggestions for the Improvement of the Hookworm Film", (n. 19), p. 1. Suggestion of "W.P.J." regarding Subtitle 7.

22. Rockefeller Archive Center, RF, Record Group 1.1, Series 100, Box 5, Folder 40, Memorandum, "Suggestions for the Improvement of the Hookworm Film", (n. 19), p. 4. Suggestion of "Hampton" regarding Subtitle 43.

23. Rockefeller Archive Center, RF, Record Group 1.1, Series 100, Box 5, Folder 40, Memorandum, "Suggestions for the Improvement of the Hookworm Film", (n. 19), p. 4. Suggestion of W.C. Hausher regarding Subtitle 44.

24. Rockefeller Archive Center, RF, Record Group 1, Series 100, Box 5, Folder 43, bound volume, Memorandum, "Malaria Film", (n. 14), p. 2.

25. Rockefeller Archive Center, RF, Record Group 1, Series 100, Box 5, Folder 39, Memo, Distribution of films, 23 December 1924. Sales of this film were most brisk in the first year of distribution, 1921, when thirty-six copies were sold. After another solid year in 1922 (twenty-nine copies sold), sales fell sharply in 1923 (thirteen copies) before rebounding in 1924 (nineteen copies).

26. Rockefeller Archive Center, RF, Record Group 1, Series 100, Box 5, Folder 39, Memo, Distribution of films, 23 December 1924, English-Domestic (n. 25). Strangely enough, in addition to the majority of films which were sent to various U.S. "domestic" states, this list includes films sent to: Geneva, Switzerland; Bangkok, Siam; Jamaica; Peking, China; Rio de Janeiro, Brazil; London, England; Dutch Guiana; and two to Montreal, Canada. There is another page with additional sales made between 23 June and 2 December 1924. These copies were all sent to either New York or Kentucky except one copy which was sent to Shanghai, China. Based on the destinations, I have added these copies to the English-Domestic tally for 1924.

27. Rockefeller Archive Center, RF, Record Group 1, Series 100, Box 5, Folder 39, Memo, Distribution of films, 23 December 1924, English-Foreign, (n. 25). Single copies were distributed widely across the globe: Melbourne and Brisbane, Australia; Madras and Calcutta, India; Porto (sic) Rico; Jesselton, Borneo; Suva, Fiji; Western Samoa; Manila, P.I.; Port Louis, Mauritius; Cairo, Egypt; and Java.

28. Rockefeller Archive Center, RF, Record Group 1, Series 100, Box 5, Folder 39, Memo, Distribution of films, 23 December 1924, Spanish (n. 25). Spanish-language copies were sent to: Quebradillas, Porto (sic) Rico; Venezuela; Managua, Nicaragua; Panama; Mexico City, Mexico; Bogota, Colombia; and New York, N.Y. . This last copy was charged to "Inter. Petroleum Co. Mdse. Acct." which is likely the International Petroleum Company, Merchandise Account.

29. Rockefeller Archive Center, RF, Record Group 1.1, Series 100, Box 5, Folder 40, Memorandum, "Unhooking the Hookworm" (n. 15), p. 2. A shipping charge of $2.50 was added "for the galvanized iron case." Based on information from the National Museum of Photography, Film & Television (web sourced information sheet 5.3.52, "Nitrate Film", NMPFT 2000), it is most likely that the 16 mm copies were produced on safety or non-flammable stock. Some of the 35 mm prints may have been produced on nitrate film which is highly flammable. The flammability risk of nitrate stock increases as the film ages and begins to decompose. The film can spontaneously ignite, even at temperatures as low as 49 C or 120 F, burning quickly and releasing toxic gases. The NMPFT informa-

tion sheet notes that all 16 mm film was produced on safety stock but that 35 mm film was made on both safety and nitrate stock.

30. Rockefeller Archive Center, RF, Record Group 1, Series 100, Box 5, Folder 43, bound volume, Memorandum, "Malaria Film" (n. 14), p. 2.

31. Rockefeller Archive Center, RF, Record Group 1.1, Series 100, Box 5, Folder 39, Letter from Rollin C. Dean, Purchasing Department, Rockefeller Foundation to Mr. T. (Thomas) G. Hull, Director, Scientific Exhibit, The American Medical Association. Type unclear on price of prints on 35mm regular stock. Could be either $85.00 or $35.00. Dean notes that the cost of the prints represents the cost of the prints to the Rockefeller Foundation.

32. Rockefeller Archive Center, RF, Record Group 1, Series 100, Box 5, Folder 43, bound volume, Memorandum, "Malaria Film" (n. 14), p. 1.

33. The author tells of the suggestion by Dr. C.-E. A. Winslow of using a 'scenario' prepared by Dr. Sella of the League of Red Cross Societies. This scenario was deemed to have, "among other defects", the deficiency of not being adapted for use outside Italy. Rockefeller Archive Center, RF, Record Group 1, Series 100, Box 5, Folder 43, bound volume, Memorandum, "Malaria Film" (n. 14), p. 1.

34. Rockefeller Archive Center, RF, Record Group 1, Series 100, Box 5, Folder 43, bound volume, Memorandum, "Malaria Film" (n. 14), p. 1.

35. Rockefeller Archive Center, RF, Record Group 1, Series 100, Box 5, Folder 43, bound volume, Memorandum, "Malaria Film" (n. 14), p. 1-2.

36. Rockefeller Archive Center, RF, Record Group 1, Series 100, Box 5, Folder 43, bound volume, Memorandum, "Malaria Film" (n. 14), p. 2. The major difference between the two films was that the hookworm film took up a great deal of staff members' time. The malaria film would involve "only general supervision of the staff".

37. Rockefeller Archive Center, RF, Record Group 1, Series 100, Box 5, Folder 43, bound volume, Contract/Letter, 7 October 1922, from Edward Stuart to C.C. Williamson, pp. 1-2.

38. Rockefeller Archive Center, RF, Record Group 1, Series 100, Box 5, Folder 43, bound volume, "Memorandum Concerning the Malaria Film", 19 October 1922, p. 2. Williamson believed that additional language titles should cost no more than $100, not $500 as Stuart had proposed. On the point of cost of prints, Williamson suggested $125, well below the original $200 or the $170 cited in the draft contract. Having learned from the hookworm film, Williamson suggested avoiding "the expensive hand colored work on each film" to keep costs down and lowering the daily rental cost from $6 to $3 because "a rental of $6 a day would seem to be prohibitive in many places where the film is needed and might react unfavorably on the Board". Finally, also as was the case with *Unhooking the Hookworm*, "to the public, the Board would be known as both producer and distributor" although in practice the Board would only take orders and then pass them on to the Society to be filled.

39. Rockefeller Archive Center, RF, Record Group 1 A83, Series 100, Films 1941-45, Box R 1996, Minutes—Entry 37126, 1 October 1937.

40. Rockefeller Archive Center, RF, Record Group 1, Series 100, Box 5, Folder 43, bound volume, Letter, Edwin O. Jordan to Edward Stuart, 8 September 1922.

41. Rockefeller Archive Center, RF, Record Group 1, Series 100, Box 5, Folder 43, bound volume, Letter, Edwin O. Jordan to Edward Stuart, 8 September 1922 (n. 40). Jordan wrote that, "the number of malaria patients in the United States is probably underestimated. In the Public Health Reports, March 21, 1919, it is estimated that the number

of cases of malaria occurring in the United States is about 6 to 7 million annually. I prefer, however, a conservative statement." Suggestions were not always acted upon. In the final edit, the decision was made to keep the figure at one million.

42. Rockefeller Archive Center, RF, Record Group 1, Series 100, Box 5, Folder 43, bound volume, Letter, Edwin O. Jordan to Edward Stuart, 8 September 1922 (n. 40). Jordan cites an article by H. F. Gray in the 24 May 1919 issue of the *Journal of the American Medical Association.*

43. Nine physicians and professors came from American states (Alabama—Dr. Kenneth F. Maxcy, State Board of Health; Louisiana—Prof. C.C. Bass; Mississippi—Dr. H.H. Howard and Dr. W.S. Leathers, State Department of Health; North Carolina—Dr. H.A. Taylor, County Department of Health; Tennessee—Dr. L.D. Fricks and Dr. M.A. Barber; Virginia—Dr. T.H.D. Griffitts, U.S. Public Health Service; Washington, D.C.—Dr. H.R. Carter) Dr. van Wesep's address is unknown. Copies were also posted to Dr. Mark F. Boyd in Rio de Janeiro, Brazil and to Dr. J.A. LePrince in Tampico, Mexico.

44. Rockefeller Archive Center, RF, Record Group 1, Series 100, Box 5, Folder 43, bound volume, "Suggested Names to receive Malaria Scenario (3rd draft).

45. Rockefeller Archive Center, RF, Record Group 1, Series 100, Box 5, Folder 43, bound volume, "Malaria Scenario (3rd draft)" (n. 44). This gives a list of forty "Numbered copies assigned as follows" and includes handwritten and typewritten notes on when marked-up copies were returned. Among the Foundation staff who were sent copies were Drs. Rose, Russell and Gregg, likely Wickliffe Rose, Paul F. Russell and Alan Gregg.

46. Rockefeller Archive Center, RF, Record Group 1, Series 100, Box 5, Folder 43, bound volume, "A.G. Suggestions." "A.G." is the only hint as to the author of this one-page memorandum. I am making the educated guess that the author of these suggestions is Alan Gregg for three reasons: I have found no other individual, in the lists of those who were asked to review scenarios, with these initials; Gregg was a member of the Foundation staff at that time; and because it was common practice at the Rockefeller Foundation for officers to sign documents with only their initials.

47. After completing his medical training at Harvard Medical School, Gregg joined the Massachusetts General Hospital. He spent time with the Harvard Medical Unit treating casualties of war in France and it was upon his return to the U.S. that Gregg began his career as a Field Staff Member with the International Health Board (1919-1922). That first position with the IHB had Gregg working on hookworm disease in Brazil. From Theodore M. Brown, "Alan Gregg and the Rockefeller Foundation's Support of Franz Alexander's Psychosomatic Research," *Bulletin of the History of Medicine* (1987) 61:158-159.

48. Rockefeller Archive Center, RF, Record Group 1, Series 100, Box 5, Folder 43, bound volume, "Scenario—Stat. Copy 13.381 SO 2". Although I believe this seventeen-page draft to be E. O. Jordan's marked-up copy, the titles and page numbers correspond to the A.G. memo. The RAC also has complete third and fifth drafts of the scenarios. The third draft has number 25 on the cover sheet. It went out to "Dr. Smillie".

49. Rockefeller Archive Center, RF, Record Group 1, Series 100, Box 5, Folder 43, bound volume, "A.G. Suggestions" (n. 46). The underlining is A.G.'s.

50. Rockefeller Archive Center, RF, Record Group 1, Series 100, Box 5, Folder 43, bound volume, "A.G. Suggestions" (n. 46). This was in reference to Page 4, Scene 14.

51. Rockefeller Archive Center, RF, Record Group 1, Series 100, Box 5, Folder 43, bound volume, "Scenario—Stat. Copy 13.381 SO 2", page 16.

52. Rockefeller Archive Center, RF, Record Group 1, Series 100, Box 5, Folder 43, bound volume, "A.G. Suggestions" (n. 46). This point has a number of check marks beside it and handwritten, "agree F.F.R." beside it.

53. Rockefeller Archive Center, RF, Record Group 1, Series 100, Box 5, Folder 43, bound volume, "Malaria: An Educational Motion Picture", 1925, p. 10.

54. Dr. Samuel T. Darling was the Director of a Rockefeller Foundation-sponsored malaria field station in Leesburg, Georgia between 1922 and 1925. He was killed in an automobile accident near Beirut while on a special mission for the malaria commission of the Health Section of the League of Nations on 21 May 1925, aged 52.

55. Rockefeller Archive Center, RF, Record Group 1, Series 100, Box 5, Folder 43, bound volume, "A.G. Suggestions" (n. 46). This is in reference to the draft copy's Scene 60, Page 16.

56. Rockefeller Archive Center, RF, Record Group 1, Series 100, Box 5, Folder 43, bound volume, "A.G. Suggestions" (n. 46). This point has "do not agree F.F.R" beside it".

57. Rockefeller Archive Center, RF, Record Group 1, Series 100, Box 5, Folder 43, bound volume, "Malaria: An Educational Motion Picture", 1925, p. 11.

58. Likely Dr. Paul F. Russell, a Rockefeller Foundation-supported malariologist.

59. Rockefeller Archive Center, RF, Record Group 1, Series 100, Box 5, Folder 39, Letter, C.C. Williamson to Edward L. Stuart, 16 April 1923. Williamson also used this opportunity to get material to update *Unhooking the Hookworm.* He wrote, "We are, as you know, very anxious to replace the 'stills' in the present hookworm film by some good moving pictures that will fit in properly with the rest of the film. If you do succeed in getting something satisfactory, I have little doubt that the International Health Board will be glad to purchase it from you."

60. Rockefeller Archive Center, RF, Record Group 1, Series 100, Box 5, Folder 39, Letter, C.C. Williamson to Professor H.C. Heaton, 18 April 1923. In an internal memo dated "4/30", there are drafts of titles translated into French, Italian, Portuguese, and Spanish (see Rockefeller Archive Center, RF, Record Group 1(.1), Series 100, Box 5, Folder 44, Correspondence).

61. Rockefeller Archive Center, RF, Record Group 1, Series 100, Box 5, Folder 39, Letter, C.C. Williamson to Professor H.C. Heaton, 18 April 1923 (n. 60).

62. Rockefeller Archive Center, RF, Record Group 1, Series 100, Box 5, Folder 39, Letter, C.C. Williamson to E. L. Stuart, 4 August 1923.

63. Rockefeller Archive Center, RF, Record Group 1, Series 100, Box 5, Folder 39, cataloguing card, entry 2316, signed by Thorvald Solberg, Register of Copyrights. This card gives Stone's address as P.O. Box 591, Carmel, California and the date of "publication" as 1 November 1922 and "2 copies each received" 30 August 1923.

64. Rockefeller Archive Center, RF, Record Group 1, Series 100, Box 5, Folder 39, Assignment of Copyright/Bill of Sale, 16 April 1924, signed by George E. Stone.

65. Rockefeller Archive Center, RF, Record Group 1, Series 100, Box 5, Folder 39, Certificate, 16 April 1924, application made by George E. Stone and signed by J.E. Beck, Notary Public in and for the County of Monterey and State of California.

66. Rockefeller Archive Center, RF, Record Group 1, Series 100, Box 5, Folder 39, Certificate from the Copyright Office of the United States of America, Library of Congress—Washington, seal affixed 13 May 1924 and signed by William Brown, Assistant Registrar of Copyrights. The copyright was recorded in the Copyright Office, book 119, page 64.

67. There is nothing on Dr. William B. Herms in the Rockefeller Foundation locator cards, the Foundation biography file or the IHD cards.

68. Martin S. Pernick offers some insight into Dr. E.H. Kleinschmidt in "Thomas Edison's TB Films: Mass Media and Health Propaganda", *Hastings Centre Reports*, (June 1978), 24. Pernick also told me that Kleinschmidt wrote several articles on early health films. Kleinschmidt died in 1960.

69. Rockefeller Archive Center, RF, Record Group 1, Series 100, Box 5, Folder 39, cataloguing card, entry 2728, also signed by Thorvald Solberg, Register of Copyrights. This card lists "Malaria Scenario by" Darling, Herms, Kleinschmidt and Stuart but follows with, "Motion picture by International Health Board of the Rockefeller Foundation, of the United States. 3 Reels". The date of "publication" is given as 12 July 1924 and "2 copies each received" 29 September 1924.

70. Rockefeller Archive Center, RF, Record Group 1, Series 100, Box 5, Folder 43, bound volume, "Malaria: An Educational Motion Picture", 1925 (n. 57).

71. Rockefeller Archive Center, RF, Record Group 1, Series 100, Box 5, Folder 43, bound volume, "Malaria: An Educational Motion Picture", 1925 (n. 57).

72. Identified, in the Rockefeller Archive Center IHD card set, as Hector H. Howard. No other information given. Thanks to Michele Hiltzik at the RAC for checking this.

73. Rockefeller Archive Center, RF, Record Group 1.1, Series 100, Box 5, Folder 39, Letter, 8 October 1931, Dr. Walter C. Earle to Dr. H. H. Howard. The film included specific location shots and local Puerto Rican health inspectors, people and businesses coupled with old Foundation film animation footage of parasites in the mosquito's gut, p. 2 of accompanying titles. The film ends with the sixty-ninth title, "With financial cooperation from all concerned, property owner, municipal governments and insular government little by little its ravages can be reduced to a minimum" and a view of the Alcaldia Vega Baja, Fajardo Sugar Company—Capitolio, p. 11 of titles.

74. Rockefeller Archive Center, RF, Record Group 1.1, Series 100, Box 5, Folder 39, Letter, 14 October 1931, Dr. H. H. Howard to Dr. Walter C. Earle. Howard does disagree with one of the titles (number 22) wherein it reads, "quinine will not prevent the infection." Howard argues this is true: "but I feel that it is not true in all of its implications. If one exposed to malaria kept a sufficient quantity of quinine in the blood current to all practical purposes the infection would be prevented if we interpret the word 'infection' to mean the development of the malaria parasites in the blood. Of course, it has been shown in Italy and elsewhere that it is not practicable to protect a population against malaria with quinine prophylaxis."

75. Rockefeller Archive Center, RF, Record Group 1.1, Series 100, Box 5, Folder 39, Letter, 6 November 1936, Rollin C. Dean to Dr. J.N. Baker, pp. 1-2.

76. Rockefeller Archive Center, RF, Record Group 1.1, Series 100, Box 5, Folder 39, Letter, 30 July 1931, Rollin C. Dean to American Medical Association, p. 1.

77. Rockefeller Archive Center, RF, Record Group 1.1, Series 100, Box 5, Folder 39, Letter, 28 August 1931, Rollin C. Dean to American Medical Association, p. 1. Positive prints of *Malaria* could be purchased for $72.00 (professional stock), $100.00 (Safety stock) or $21.40 (16mm). Prints of *Unhooking the Hookworm* could be purchased for $35.00 (professional or Safety stock) or $8.25 (16 mm format).

78. Rockefeller Archive Center, RF, Record Group 1.1, Series 100, Box 5, Folder 39, Letter, 13 April 1936, Dr. Mark F. Boyd to Dr. John A. Ferrell, p.1.

79. Rockefeller Archive Center, RF, Record Group 1.1, Series 100, Box 5, Folder 39, Letter, 13 April 1936, Dr. Mark F. Boyd to Dr. John A. Ferrell (n. 78), pp. 1-2.

80. Rockefeller Archive Center, RF, Record Group 1.1, Series 100, Box 5, Folder 39, Letter, 13 April 1936, Dr. Mark F. Boyd to Dr. John A. Ferrell (n. 78), p. 2. He cites Earle's "very good two reel film" (p. 2) which used selections from the Foundation film, and other new films including ones made by the Bureau of Entomology and Public Health Service.

81. Rockefeller Archive Center, RF, Record Group 1.1, Series 100, Box 5, Folder 39, Letter, 13 April 1936, Dr. Mark F. Boyd to Dr. John A. Ferrell (n. 78), p. 2.

82. Rockefeller Archive Center, RF, Record Group 1.1, Series 100, Box 5, Folder 39, Letter, 13 April 1936, Dr. Mark F. Boyd to Dr. John A. Ferrell (n. 78), p. 2-3. There is a handwritten note beside this paragraph that reads, "Used to be done quite widely (?) even w/ 35 m/m". The note is written by the final Foundation officer who initialled having read the letter and I believe the initials read "RCD," likely Rollin C. Dean.

83. Rockefeller Archive Center, RF, Record Group 1.1, Series 100, Box 5, Folder 39, Minutes, 20 May 1936. This appears to be an internal Foundation memo. Only initials are used for those present and to record their comments. The list of those who were at this meeting (with likely identities in parentheses, thanks to Michele Hiltzik at the RAC) is given as: JAF (John A. Ferrell, Foundation officer), AJW (Andrew J. Warren, IHD fellowship advisor, 1928-1937), LTC (Dr. Lowell T. Coggeshall, malariologist and RF special staff member), DFM (Daniel Franklin Milam, 1935-1937, "assigned to Foundation's NY laboratories where clinical trials were being conducted on Malaria fever", from RF biography file, RCD (Rollin C. Dean, RF staff), HM (likeliest Dr. Hugo Muench, Jr., staff 1935-1946), PP (Persis Putnam, IHD statistician, 1926-1948), and RVR (unidentifiable). JAF (Ferrell) summarizes Boyd's comments regarding the need for more than one film and to include "local color" (p. 1). 'HM' (Muench) agrees with adding local pictures. The cost of the film was given as "Five or six thousand dollars", the cost of supplementing existing footage "without introducing individuals in it" as "$80 to $90", and the age of the film as "nearly fifteen years old" (p. 2).

84. Rockefeller Archive Center, RF, Record Group 1.1, Series 100, Box 5, Folder 39, Minutes, 20 May 1936, p. 4 (n. 83).

85. Rockefeller Archive Center, RF, Record Group 1.1, Series 100, Box 5, Folder 39, Minutes, 20 May 1936, p. 4 (n. 83). 'PP' pointed out that the Foundation had not only the information, research and facts to make the most up-to-date films, but also "opportunity to produce something of value and general information" but 'RCD' argued that "Films were never developed for that purpose, nor for wide distribution."

86. Rockefeller Archive Center, RF, Record Group 1.1, Series 100, Box 5, Folder 39, Minutes, 20 May 1936, p. 4 (n. 83).

87. Rockefeller Archive Center, RF, Record Group 1.1, Series 100, Box 5, Folder 39, Letter, 6 August 1937, W.A. Sawyer to Surgeon General.

88. Rockefeller Archive Center, RF, Record Group 1.1, Series 100, Box 5, Folder 39, Letter, 6 August 1937, W.A. Sawyer to Surgeon General (n. 87). Sawyer concludes with the statement that the Foundation would "naturally prefer" to give the films to a public health agency, rather than "to give or sell them to a commercial organization".

89. Rockefeller Archive Center, RF, Record Group 1 A83, Series 100, Films 1941-45, Box R 1996, Minutes—Entry 37126, 1 October 1937. The Minutes record the donation of the negatives of the two films, as well as additional negatives "of scenes and titles which are related to these films, although not actually included." The Minutes also outline the situation—the films had become outdated but segments might still prove useful—and cited the initial cost of *Malaria* as $10,523 and of *Unhooking the Hookworm* at

$4,641. By late 1937, both films were worth no more than $1,400. Rockefeller Archive Center, RF, Record Group 1.1, Series 100, Box 5, Folder 39, Letter, 24 November 1937, Robert Olesen, Assistant Surgeon General to Dr. Wilbur A. Sawyer, Director, International Health Division, Rockefeller Foundation. Olesen acknowledges receipt of eight reels of *Malaria* and three reels of *Unhooking the Hookworm*.

90. Rockefeller Archive Center, RF, Record Group 1.1, Series 100, Box 5, Folder 39, Letter and attached correspondence, 26 April 1938, W.P. Jacocks, International Health Division representative in India, Rockefeller Foundation to Dr. W.A. Sawyer.

91. Rockefeller Archive Center, RF, Record Group 1.1, Series 100, Box 5, Folder 39, Letter, 25 May 1938, likely Dr. W.A. Sawyer (signature obscured) to Dr. W.P. Jacocks. There is also the example of the Georgia Department of Health who, in 1938, requested some footage from *Unhooking the Hookworm* which they would use in a new, revised and re-edited public health film (Rockefeller Archive Center, RF, Record Group 1.1, Series 100, Box 5, Folder 39, Letter, 11 October 1938, M.V. Ziegler, Regional Consultant, U.S. Public Health Service to Dr. John A. Ferrell).

92. Rockefeller Archive Center, RF, Record Group 1.1, Series 100, Box 5, Folder 39, Letter, 5 August 1938, J.T. Alexander to The Rockefeller Foundation.

93. Rockefeller Archive Center, RF, Record Group 1.1, Series 100, Box 5, Folder 39, Letter, 18 August 1938, W.A. Sawyer to J.T. Alexander.

94. Rockefeller Archive Center, RF, Record Group 1.1, Series 100, Box 5, Folder 39, Letter, 18 August 1938, W.A. Sawyer to J.T. Alexander (n. 93). Copies of *Malaria* were still being requested as late as 1941 by public health officials in Argentina (see Rockefeller Archive Center, RF, Record Group 1 A83, Series 100, Films 1941-45, Box R 1996, Letter, Justin Andrews to Dr. L. W. Hackett and correspondence between Justin Andrews and Rollin C. Dean.

95. Rockefeller Archive Center, RF, Record Group 1.1, Series 100, Box 5, Folder 39, Memorandum, "Visual Education", 21 December 1936, p. 1.

96. Rockefeller Archive Center, RF, Record Group 1.1, Series 100, Box 5, Folder 39, Memorandum, "Visual Education", 21 December 1936, p. 2 (n. 95). The hookworm film was purchased by twenty-one countries.

97. Rockefeller Archive Center, RF, Record Group 1.1, Series 100, Box 5, Folder 39, Memorandum, "Visual Education", 21 December 1936, p. 3 (n. 95).

98. Copies of this animated film were requested by public health officials as far away as British Guiana (see correspondence between Dr. Porter J. Crawford and Dr. George Bevier, Rockefeller Archive Center, RF, Record Group 1 A83, Series 100, Films 1941-45, Box R 1996).

99. *Winged Scourge*, including an image of one of the dwarves screwing a screen on a door, was mentioned in a brief article on the importance of screens in keeping mosquitoes out in TIME magazine, 14 June 1943, pp. 46-47. A copy of this article can be found in Rockefeller Archive Center, RF, Record Group 1.1, Series 100, Box 47, Folder 460, "Screen Salesman".

100. This film was part of the 'Health to the Americas Series'. All available in Spanish, the series included: *La Limpieza Trae Buena Salud* (*Cleanliness Brings Good Health*), an eight-minute, twenty second short released in 1944; *Le Enfermedad se Propaga* (*How Disease Spreads)*, a ten-minute, seventeen second film also released in 1944; and *Saneamiento del Ambiente* (*Environmental Sanitation*), an eight-minute, twenty-nine second short released in 1945.

101. There was an earlier film on malaria in Italy, proposed by Prof. B. Gosio (Rockefeller Archive Center, RF, Record Group 1, Series 100, Box 5, Folder 39, "Outline of Proposed Italian Film on Malaria", 1925. The single-page outline covered the damage done by malaria including depopulation; the history of "supposed causes of malaria"; modern scientific understanding (using material from *Malaria*); and combative measures, including quinine, "training of the army which will fight malaria: School at Nettuno", 'Grande Bonifica' canals and pumps, drainage and filling, Paris Green, mosquito-eating fish, and underdrainage.

102. The film was produced by the Ministry of Education and Health in Brazil, Service of Studies and Investigations in co-operation with the Foundation and outlined, documentary-style, the anti-gambiae campaign in Brazil. It showed how Rockefeller Foundation initiatives, mainly relying on spraying the arsenic-based compound Paris Green, helped to quell the spread of malaria in northeastern Brazil. No doubt because of increased wartime interest in malaria prevention, it was part of a special midday screening, along with *Winged Scourge* and *Jungle Yellow Fever,* for all Foundation officers at the Museum of Modern Art on 10 September 1943. (from Rockefeller Archive Center, RF, Record Group 1 A83, Series 100, Films 1941-45, Box R 1996, Memo, 26 August 1943, from WAS (Wilbur A. Sawyer) to "RF Officers, including FBH and HMM".

103. Rockefeller Archive Center, RF, Record Group 1 A83, Series 100, Films 1941-45, Box R 1996, "Films of IHD Activities", February 1950, p. 2—"Films in I.H.D. Storeroom, November 1949". The footage included material filmed by Foundation officers Payne, Hackett, Knipe and Soper. The malaria footage dates between 1924 and 1931 and covers the United States, Argentina, Puerto Rico, Italy, Albania, and Greece.

104. H.B. van Wesep, based on information from Rockefeller Foundation locator cards, began his employment with the Foundation on 24 April 1916. He served as Assistant to the Chief of Information Services from 1921 to 1926 when he took over as Chief of Information Services. He retired in 1953, one year after this letter was written.

105. Rockefeller Archive Center, RF, Record Group 1 A83, Series 100, Films 1941-45, Box R 1996, Letter, 11 January 1952, H.B. van Wesep, Office of Publications to Homer J. Dietmeier, Boston University. The Foundation would continue to participate in filmmaking on a sporadic basis throughout the 1970s. See list of recent acquisitions, Rockefeller Archive Center, RF, Record Group 1 A83, Series 100, Films 1941-45, Box R 1996, "Film Collection at Archives", 20 May 1975, most of which deal with India, Albania, Sardinia, Florida, Mexico, Thailand, and outtakes from existing films.

Chapter Three

Foundations, Eugenic Sterilization, and the Emergence of the World Population Control Movement

David A. Valone

As academics, we often toil in isolation and relative obscurity. This isolation, in all honesty, is in part the product of our own unwillingness or inability to bring the fruits of our labor to a wider audience. Another contributing factor is the somewhat self-defeating tendency we have to scorn those amongst us who do achieve public notoriety. This situation is unfortunate in many ways, not the least of which is the fact that the work we do, as is hopefully the case with this book, can potentially be both of great practical value and of deep significance to issues of wide importance. One sad result of this academic exile was brought home to me with particular force in the summer of 2003, when a research interest of mine suddenly became the hot news issue of the day. At that time, the Senate of the California State legislature held hearings regarding the history of that state's aggressive sterilization laws. The state Senate called upon a number of experts, most notably Paul Lombardo of the University of Virginia's Bioethics Institute. When the Senators heard Lombardo's grim testimony concerning the roughly 20,000 people who were subjected to state sponsored sterilization during the previous century, they reacted with shock and horror. Most Californians, and apparently all of the members of the Senate panel, had no idea that these sterilizations had occurred; later that day, the state's Governor Gray Davis issued a sweeping apology on behalf of the state to those who had been affected.

So what is my connection to all this? Let me quote briefly from the opening of an article by reporter Mike Anton on the fallout from those hearings that appeared on the front page of the *Los Angeles Times* under the headline "Forced Sterilization Once Seen as a Path to a Better World:"

> In a basement at Caltech, 59 gray boxes contain thousands of documents that reveal in detail how an influential group of California men once hoped to direct the fate of the human race. Within the brittle files is the story of the state's long and largely forgotten effort to sterilize mental patients. Memos show how civic leaders helped popularize eugenics around the world, including Nazi Germany. Case histories offer a glimpse of the more than 20,000 people who were, by law, sterilized in state hospitals from 1909 through the 1960s. . . . The Caltech archive contains 16 boxes of patient cases that will remain closed until 2005. But hundreds of unnamed records offer sketchy profiles of patients and the motives of doctors who operated on them.[1]

With a slight tip of the hat to Anton's dramatic prose, let me say that I know those fifty-nine gray boxes in the basement at Caltech very well, and I'm one of only two or three people who have looked through them all. The documents, by the way, are not that brittle, with the exception of the several hundred individual case reports on people who were sterilized at the Sonoma State Home in Eldridge, California. Those are the records in the boxes that remain closed—they are still closed, and will remain so until at least 2010. During the early 1990s, I worked for several years as the staff historian in the archives of the California Institute of Technology. At that time, I was presented with a fascinating opportunity to process the papers of the Human Betterment Foundation. The papers had been discovered in a Caltech warehouse in 1968 where they apparently had been stored since the dissolution of the Human Betterment Foundation in the early 1940s. The papers, however, remained closed to researchers due to legal issues relating to personal medical records in the collection. By the early 1990s, a decision was made by the Caltech Archivist, Judith R. Goodstein, to process the collection and open it to researchers. The task of processing the papers fell to me, and the collection was opened for researcher after I finished the project in 1995.

The question, of course, is this: why should the existence of an archival collection available for nearly a decade—and for that matter the existence of the California sterilization program—suddenly become a hot political issue and front-page news in the *Los Angeles Times*? The answer to this question, without much doubt, lies both in the nature of political discussions in this country and the sensationalistic nature of the journalistic enterprise. Beyond these somewhat intractable issues, though, rests a very real and important concern about the way that aspects of the past are suppressed, forgotten, or conveniently written out of our history. By this I mean both our grand, national history and the more localized, and perhaps even more influential, historical memories of particular enterprises and institutions. In thinking about how the past matters, as we are doing in this book, we should not forget that people in positions of influence and power make decisions everyday that they consider informed judgments based on past experiences. Those experiences, though, are limited by what gets remembered as history and what gets forgotten.

Of course the ironic thing about all the uproar in California is that there really is not anything particularly new in the story of the state's sterilization program that "broke" at those hearings, at least to those who work in the history of eugenics. The general outlines of the eugenics movement and in the U.S., Britain, and Germany have been well established during the past two decades by several major historical monographs. The role of philanthropic foundations, particularly the Rockefeller Foundation, in funding eugenics both in the United States and around the world, has also been well documented.[2] There are aspects of these stories, however, that have yet to be told. One recent trend has been studies of eugenics movements in various national or regional contexts. Of particular relevance for us are the studies that examined eugenic ideologies as an aspect of western imperial domination in the twentieth century, most notably Laura Briggs' recent work on the ideological dimensions of U.S. scientific intervention in Puerto Rico.[3] Comparative studies, as well as investigations of the links among various national movements, are also shedding new light on both the social and scientific underpinnings of eugenics. There have also been recent calls for a reevaluation of aspects of the standard historical model of the development of the movement, which include both an effort to broaden the notion of "eugenics" under a more encompassing model of social control and a reconsideration of the purely hereditarian cast of efforts at eugenic reform. A detailed history of the Human Betterment Foundation will someday be an important part of this larger story of eugenics.[4] The history of the Human Betterment Foundation also bears additional scrutiny in the context that we are discussing in this book: the role of philanthropic foundations in the globalization of western medicine and public health. In particular, I want to discuss the Human Betterment Foundation's role as a worldwide promoter of eugenic sterilization, and its association with a successor organization, the Human Betterment Association for Voluntary Sterilization.

To begin this discussion, let me provide a brief history of the Human Betterment Foundation (HBF), a non-profit organization chartered in 1929. The goal of the Foundation was the promotion of eugenic sterilization, an ideal that was put into practice primarily through the distribution of books and pamphlets including detailed case studies drawn from the sterilizations of those judged mentally defective under the California sterilization laws passed in 1909. E. S. Gosney was the founder and President of the HBF. He was born in 1855, graduated from Richmond College in Missouri in 1877, and received a law degree from the St. Louis School of Law three years later. Gosney eventually settled in the territory of Arizona. There he set up a successful law practice in Flagstaff, and also became involved in the financing of a number of businesses, particularly in the sheep and cattle breeding industry. He organized the Arizona Wool Growers' Association to fight for the rights of small stock farmers who faced elimination at the hand of land speculators and railroad companies.[5] Around 1905, Gosney began spending his winters in Pasadena and soon decided to relocate to southern California, in part to provide a more refined environment for the education of his two daughters. He quickly became a leading member of the Pasadena business community, buying up citrus fields and real estate in the still sparsely populated San Gabriel valley just outside of Los Angeles. In 1906, he became the principle financier and chairman of the board of trustees of the Polytechnic Elementary School, an institution dedicated to providing high quality elementary education. Gosney also became very active in the leadership of the California branch of the Boy Scouts of America.

While in Pasadena, Gosney became a close associate of Paul Popenoe, a very significant figure in the history of American eugenics and the internationalization of the eugenic ideology. Together, Popenoe and Gosney began an extended study of the medical, legal, and social aspects of the sterilizations being carried out under the terms of the California Sterilization Laws at the Sonoma State Home and other state institutions. The result of this work, entitled "Sterilization for Human Betterment," was published in 1929. In that same year, Gosney set up the HBF and gathered a membership of twenty-five leading California scientists and philanthropists including Popenoe, as well as Lewis Terman, David Starr Jordan, William B. Munro, Otis Castle and C.M. Goethe.

During the next thirteen years, the HBF continued to carry out research on the effects of sterilization and undertook widespread distribution of their pamphlet "Sterilization for Human Betterment" to individuals, public libraries, and schools both in the United States and around the world. During this period, ties between the HBF and its Pasadena neighbor, the California Institute of Technology, also began to grow. Caltech's president, Robert Millikan, shared aspects of Gosney's vision for human progress; he also had a keen eye for potential donors to Caltech. Millikan joined the board of directors of the HBF in 1937. Gosney also courted Thomas Hunt Morgan, then director of Caltech's Division of Biological Sciences, for Morgan's support for the Foundation during the early 1940s.[6]

Upon Gosney's death in 1942 his daughter, Lois Gosney Castle assumed the leadership of the Foundation. Together with the HBF's Board of Trustees, she decided to liquidate the assets of the Foundation and turn the proceeds over to Caltech. In 1943, an agreement was drawn up between the HBF and Caltech, wherein Caltech agreed to use the Foundation's assets to set up a research fund in memory of E.S. Gosney that would be administered by Caltech's Division of Biology. This fellowship, intended to carry on the spirit of the Foundation's work for human betterment, has been used to support post-doctoral research "in those branches of biological science basic to our understanding of human welfare."[7]

So the Human Betterment Foundation existed for only the brief span of fourteen years, only the blink of an eye in comparison to the tenure of many philanthropic foundations. And yet during its day the HBF played an extremely critical role both in the promotion of the sterilization laws already on the books in California, and in the growing international interest in sterilization as a means of eugenic improvement and public health. In this regard, Paul Popenoe was the key player. Based on the Archival records of the HBF, it is clear that while Gosney may have been the financial and ideological inspiration for the Human Betterment Foundation, it was Popenoe who was its main voice and its primary promoter both in the United States and internationally. Popenoe was a tireless advocate of eugenics during the 1920s and the 1930s, and he corresponded extensively with leaders of the American eugenics movement and with a dizzying number of international admirers. Letters in the HBF archives to the Foundation and to Popenoe come from every area of the world, including Africa, Asia Minor, Central and South America, China, Russia, and Southeast Asia, and the Pacific Islands. While neither Popenoe nor Gosney had any medical training, the efforts of the Human Betterment Foundation were especially directed toward doctors. The Foundation sponsored research on the effects both of salpingectomy (now more commonly referred to as a tubal ligation) and vasectomy, including both the physical and the psychological effects of sterilization on subjects. Their efforts were primarily focused on convincing doctors of the safety and ease of surgical sterilization. Judging from the letters to the HBF from international sources, the Foundation was effective in promoting their ideas to doctors working around the world. Many of these physicians were quite proficient in English, indicating either that they had been trained in the United States or in a colonial medical facility.[8]

In addition, the HBF had pursued very active ties with the Nazis in the early 1930s. In 1937, Dr. Fritz Lenz, a leading German medical geneticist, wrote to Gosney to thank him for sending his latest data on the California sterilization program, since "these practical experiences are also very valuable for us in Germany." A leading donor to the HBF, C. M. Goethe, visited Nazi Germany in the mid-1930s and reported to Gosney that the HBF's work "had a powerful part in shaping the opinion of the group of intellectuals who are behind Hitler . . . you have really jolted into action a great government of 60,000,000 people."[9] By the latter part of the decade, however, the HBF began to back off of direct connections with the Nazi eugenics program. In a letter dated 9 September 1940, Gosney wrote: "We have little in this country to consider in racial integrity. Germany is pushing that. We should steer clear of it lest we should be misunderstood."[10] Nonetheless, Gosney and Popenoe continued to support sterilization as a viable means of eugenics, both at home and abroad.

Although the Human Betterment Foundation was dissolved in 1943, the efforts that the Foundation had made promoting eugenic sterilization, especially among the medical community, were continued by two successor Foundations: Birthright, based in Princeton, New Jersey, and the Human Betterment Association of America, based in New York City. This latter organization has gone through a series of subtle and not so subtle name changes, first to the Human Betterment Association for Voluntary Sterilization (HBAVS), then the Association for Voluntary Sterilization, Inc.; it is still in operation, based in New York City and operating under the name EngenderHealth (an organization that now has as its slogan "Improving Women's Health Worldwide"). Birthright received most of the records of the Human Betterment Foundation regarding its work on sterilization programs on the United States. Birthright also continued to receive the financial and spiritual support of C.M. Goethe. The Human Betterment Association for Voluntary Sterilization took up the mission of the HBF in promoting sterilization internationally. Among its sponsors were Lois Gosney Castle, Goethe, and Popenoe.

This reorganization of the HBF in the early 1940s was part of a larger trend among eugenic organizations at the time. In the middle of the 1940s and with increased urgency in the years immediately after the war, the American eugenics movement reformulated itself along two fronts: first, as promoters of a reductionist biological science that would eventually help to foster the origins of modern molecular biology; and second, as promoters of a new worldwide population control movement. In a letter dated 31 May 1949, Columbia University professor Maurice A. Bigelow, a past president of the American Eugenics Society, wrote to Popenoe concerning the expansion of eugenic sterilization around the world since the end of WWII. He begins his letter by noting "I just received your big letter of May 14 containing articles on various countries. I think we can use all of them in the September-December 1948 [issue of *Eugenical News*] which I am getting ready at the printer. I have the page proofs for the March-June, 1948, including the article on Japanese sterilization." Later in the letter he noted that "Osborn, Lorimer, and Kiser and possibly other population men will furnish material for the March 1949 *Eugenical News*. They will base it on some papers which are to be read at the next meeting of the Population Association."[11] Popenoe by that time had begun writing popular articles on marriage counseling for magazines such as *Ladies Home Journal*, where he initiated the "Can This Marriage Be Saved" column that continues to run today. But he also continued his promotion of sterilization, now in the form of voluntary sterilization. He was listed as a member of the "Legal and Scientific Committee" on the letterhead of the Human Betterment Association for Voluntary Sterilization into the 1960s.

In the late 1940s, the HBAVS began a more aggressive program for the promotion of sterilization as a means of international population control. In November of 1949, H. Curtis Wood, M.D. the President of the Association wrote to Popenoe to report on the progress: "Our New York office is located in the New York Academy of Medicine; we are furnishing the Population Division of the Social Affairs Department of the United Nations with mimeograph copies of all state sterilization laws, Puerto Rico, and Japan, for use in their legal monograph to be published in 1951. In the November issue of the AMA Journal you will find an article by Dr. Clarence J. Gamble based on Birthright's publication #5, 'Sterilizations Officially Reported January 1st, 1949.'"[12] In 1963, Robert W. Laidlaw, M.D., President of the Human Betterment Association for Voluntary Sterilization opened a fund raising letter for the Association by noting that "The population of the world in mid-1963 is approximately 3,180,000,000. The increase of 185 million people *in only three years* almost equals the total population of the United States! . . . much of our activity since our founding in 1937 has been dedicated to educational programs designed to make known the benefits of voluntary sterilization and its growing acceptance throughout the world."[13]

The involvement of organizations like the HBAVS in international population control efforts became more direct in the 1970s. In 1973, the foundation, now under the name the Association for Voluntary Sterilization, Inc. organized an international conference in Geneva on voluntary sterilization around the world. By late 1973, it had begun twelve projects in San Salvador, Guatemala, Iran, Thailand, Colombia, Ecuador and Taiwan.[14] Since then the organization, now operating under the name EngenderHealth, has expanded its operations to more than ninety countries. The organization describes its ongoing activities in Bangladesh in the following terms:

> EngenderHealth has been working in Bangladesh since 1974 to improve family health and reduce overall fertility rates in Bangladesh. We work in partnership with several cooperating agencies and the National Integrated Population and Health Program (NIPHP), a state organization with the goal of reducing poverty and thereby increasing self-reliance among the people of Bangladesh.[15]

While the involvement of non-governmental organizations like EngenderHealth in the internationalization of public health in the form of voluntary sterilization may not seem remarkable, understanding the history of organizations such as this should certainly raise our awareness of the more subtle ideological and political agendas that might be pursued in the name of public health. Let me again quote EngenderHealth's recent literature, in this case on some of its domestic programs aimed at "low-income" markets:

> Our research has also enabled us to develop social marketing campaigns that are specifically targeted to local populations. For example, our work with a health facility in Colorado taught us that low-income men were not accessing services. Through our efforts, social marketing materials were developed that specifically responded to the needs expressed by this population . . . Since 1996, EngenderHealth has been providing training in social marketing techniques and technical guidance to the Young Men's Clinic (YMC), a facility in New York City's Washington Heights area . . . [we] produce[d] a series of cartoons addressing current sexual health issues aimed at both young women and men in the Washington Heights community, a predominantly Hispanic and low-income neighborhood.[16]

As I am suggesting, one need not read very deep into statements such as these to see the possibility of a continuity of aims with earlier eugenic sterilization. Of course current programs, both at home and abroad, are carried out under the banner of "voluntary" sterilization. Public education campaigns such as these, however, might cast some doubt upon the degree to which those subjected to them are actually in a position to make well informed choices. The possibility that such choices might be even less so internationally, under the supervision of frequently repressive political regimes, would hardly be surprising. Adding to these concerns, of course, is the very power of westernized medicine and its agenda of public health. When such ideas are pitched to indigenous populations under the banner of the United Nations, and buttressed by the instrumental effectiveness of western technology in many areas of life, the situation becomes all the more worrisome.

In 2002, EngenderHealth received the Population Award from the United Nations. In a statement congratulating the foundation on this award, Assistant Secretary-General for External Relations, Gillian Sorensen, delivered the following message on behalf of Kofi Annan:

> EngenderHealth, formerly known as AVSC International, was established in 1943 and initially worked to ensure that American couples could choose surgical sterilization and that these services were voluntary and safe. In 1973, EngenderHealth broadened its geographic scope to work in the developing world, where it has introduced and expanded access to safe and voluntary sterilization services in more than 90 countries. In large part due to EngenderHealth's pioneering work, sterilization is the most prevalent method of contraception worldwide.[17]

In thinking about the role of foundations in globalizing western medicine and public health, it seems only appropriate to reflect on the fact that sterilization has become the dominant form of birth control around the world. One certainly could argue whether the history of medical sterilization, and its association with eugenics and coercive state sponsored programs, is really relevant to the current practice of sterilization as a voluntary means of family planning. Without question there is a clear difference between individuals making free choices about their reproductive desires and state sponsored forced sterilization. Even if we set the true freedom of individual choices aside, however, when family planning becomes a state sponsored or an internationally encouraged form of population control, the history of eugenic sterilization should at least be acknowledged, especially when one considers the role of foundations and other NGOs in promoting these practices.

Notes

I wish to thank the Institute Archives of the California Institute of Technology for permission to quote from the E.S. Gosney Papers and the Records of the Human Betterment Foundation, and the American Heritage Center at the University of Wyoming for permission to cite from the Paul Popenoe and Medora Steedman Bass papers. Portions of this essay have been previously published as "Eugenic Science in California: The Papers of E. S. Gosney and the Human Betterment Foundation" in *The Mendel Newsletter. Archival Resources for the History of Genetics and Allied Sciences* New Series, No. 5 (February 1996), 13-15.

1. Mike Anton, "Forced Sterilizations Once Seen as a Path to a Better World; Decades of files on mental patients reveal how a group of noted Californians hoped to influence the fate of the human race." *Los Angeles Times* 16 July 2003, main news, part 1, p 1.

2. Edward J. Larson, *Sex Race, and Science: Eugenics in the Deep South* (Baltimore: Johns Hopkins University Press, 1995), 30; Stephan Kühl, *The Nazi Connection: Eugenics, American Racism, and German National Socialism* (New York: Oxford University Press, 1994), 20-21; Edwin Black, *The War Against the Weak: Eugenics and America's*

Campaign to Create a Master Race (New York: Four Walls Eight Windows Press, 2003), 294-96.

3. Laura Briggs, *Reproducing Empire: Race, Sex, Science, and U.S. Imperialism in Puerto Rico* (Berkeley: University of California, 2002).

4. Alexandra Stern's important book *Eugenic Nation: Faults and Frontiers of Better Breeding in America* (Berkeley: University of California Press, 2005) was published after this talk was delivered. It provides a fascinating account of various dimension of the story of eugenics in the west, including some discussion of the work of the HBF.

5. Information of Gosney's life can be found in the E.S. Gosney collection as well as the papers of Lois Gosney Castle Troendle, and in the archives historical files under "Gosney, Ezra Seymour (1855-1942)," in the Institute Archives of the California Institute of Technology.

6. Human Betterment Foundation Papers, Box 1.4. Institute Archives, California Institute of Technology.

7. George Wells Beadle, "The Gosney Research Fund," *Engineering and Science*, 10 (May 1947), 27.

8. Human Betterment Foundation Papers, Correspondence Files, Boxes 5-10. Institute Archives, California Institute of Technology.

9. Correspondence from the Human Betterment Foundation Papers cited in Anton, "Forced Sterilization," p. 1.

10. Human Betterment Foundation Papers, Box 1.2. Institute Archives, California Institute of Technology.

11. M. .A. Bigelow to Paul Popenoe, 31 May 1949. Popenoe Papers, Box 10. American Heritage Center, University of Wyoming. On origins of the transition from eugenics to population control, see Garland E. Allen, "Old Wine in New Bottles: From Eugenics to Population Control in the Work of Raymond Pearl," in *The Expansion of American Biology* ed. Keith R. Benson, Jane Maienschein, and Ronald Rainger (New Brunswick: Rutgers University Press, 1991), 231-261. In particular, Allen provides details of the Rockefeller Foundations early involvement with eugenics and population control in his section on "Funding and Institutional Support for Pearl's Population Studies."

12. H. Curtis Wood to Paul Popenoe, 30 November 1949, Popenoe Papers, Box 11. American Heritage Center, University of Wyoming.

13. Robert W. Laidlaw, M.D. to Medora Steedman Bass, 4 November 1963. Medora Steedman Bass Papers, Box #2. American Heritage Center, University of Wyoming.

14. Special Appeal fundraising letter from Helen Edey, M.D. Chairman of the AVS, October 1973, Medora Steedman Bass Paper, Box #3, American Heritage Center, University of Wyoming.

15. http://www.engenderhealth.org/ia/cbc/bangladesh.html. Accessed 6/1/04

16. http://www.engenderhealth.org/ia/cbc/united_states.html Accessed 6/1/04

17. United Nations Press Release dated 1/07/02. SG/SM/8296 POP/833. Available at http://www.un.org/News/Press/docs/2002/sgsm8296.doc.htm. Accessed 6/2/04.

Chapter Four

Selskar Gunn and Paul Russell of the Rockefeller Foundation: A Contrast in Styles

Socrates Litsios

Malaria is a health and social problem; it must be attacked simultaneously from both these angles. Selskar Gunn, 1883-1944

In those areas where malaria is the outstanding social and health problem, the resources of the health administration, specially augmented where necessary, should be directed chiefly towards malaria control, even if this should entail the restriction of other public health activities, until malaria is no longer of major importance. Paul Russell, 1894-1983

The epigraphs above by two senior RF officials are taken from the report of a meeting organized in 1937 by the League of Nations Health Organization.[1] Gunn, who was Vice-President of the foundation, had returned to Paris after working for five years in China where he was responsible for a major project whose aim was that of rural development. Russell was a senior malariologist for the foundation and was stationed in India at the time. Gunn's quote is taken from the report's introduction, which he wrote the night before the meeting ended. Russell's quote is taken from the malaria section of the report, for which he was responsible. This paper explores their different perspectives on the 'social,' and suggests some reasons for why Gunn's experience had so little impact on post-WWII developments.

The Evolution of Selskar Gunn's Approach to Public Health

Selskar Gunn received a degree in biology from MIT in 1905 where he studied under William Sedgwick, who at the time was considered by many to be the leading public health educator in America. In 1917 he received a certificate in public health from the Harvard/MIT School of Public Health; 1917 also saw him depart for Paris as a member of the Commission for the Prevention of Tuberculosis organized by the Rockefeller Foundation. Between 1905 and 1917 Gunn held a variety of jobs, including: Lecturer, Iowa State University; Health Officer, Orange, NJ; Associate Professor, MIT; Director of the Division of Hygiene, Massachusetts State Board of Health; and Editor of the American Journal of Public Health.

Before being requested to join the anti-TB commission, George Vincent, president of the Rockefeller Foundation, suggested to William Welch, Dean of the Johns Hopkins School of Public Health, that Gunn be appointed to the faculty. Vincent was particularly impressed with Gunn's work on popular health education, qualifying it as "a model of imagination, ingenuity and efficiency."[2] Welch was not interested in any appointment that would upset his plans to engage only a "highly specialized staff." He did not want "these men to be disturbed . . . by having to do popular work or by being part of a large organization part of which is meant to do popular work."[3] What the history of public health would have been had Hopkins' public health program aimed to support "popular work" deserves more attention than it has so far received. But that question is outside the purview of this paper.

In France Gunn was responsible for developing the educational campaign whose aim was to get the French public to adopt preventive measures against tuberculosis. The campaign used techniques drawn from American advertising such as dramatic multi-color posters that were new to the French. Hundreds of towns were visited by a traveling exhibit which featured a 42-panel exhibit, printed matter and film. Children became the focus of Gunn's campaign. They wrote short essays that addressed such questions as "What do you know about tuberculosis?," "How is it transmitted?," and "How can one avoid tuberculosis?"[4]

Gunn was assigned to Czechoslovakia in 1920 as an advisor to the Ministry of Health. Upon his return to Paris in 1922 he became European regional director of the foundation's International Health Division (IHD), a position which made him a central figure in all negotiations between the foundation and European countries as well as the newly established League of Nations Health Organization (LNHO). Gunn's interaction with European's leading public health specialists, in particular Andrija Štampar of Yugoslavia and Ludwik Rajchman, Director of the LNHO, helped further broaden his approach to public health.

In was in Yugoslavia that Gunn gained an appreciation of the importance of agriculture for public health. During his visit in February 1928, for example, he recorded Štampar's joy at his having had a positive audience with King Alexander concerning agriculture, and he further noted in his diary, "Štampar develops his philosophy, especially with regard to agriculture—public health secondary to development of agriculture . . . educational work more elemental—"Kittel Schule," short courses for peasants, good field staff, good agricultural experimental stations and laboratories."[5]

Encouraged by the foundation's "new policy in Social Sciences (SS)," adopted in 1929, Gunn pushed for the development of programs that went well beyond the traditional boundaries of public health, noting that many of the health officers with whom he worked "feel that the time has come when governments should consider the possibility of elaborating more unified programs." He was given the chance in October 1930 to present a 'cooperative IHD, SS, Public Welfare Program' to the Foundation's Board of Trustees and officers.

In his proposal Gunn referred to the work of the IHD in "certain European countries," out of which there was a "growing feeling that the time has come when governments should consider the possibility of elaborating more unified programs." In particular, "health officers feel that the development of a health program is an isolated effort (and as) important as such a program is, it is far from solving the complicated problems of the community." A wider effort was needed, as suggested by the initiative of some countries to create ministries of Public Welfare.[6]

Gunn illustrated the "beneficial results" that could be expected from this new initiative by an example drawn from Yugoslavia. There, it was found that epidemics of diseases in cattle were so severe as to cause great financial loss, so much so that there were not adequate funds for public health work. In such a situation "steps must be taken to introduce the known methods of preventive and veterinary medicine."[7] Gunn had been working with Štampar to develop such a new program in Yugoslavia. This new program for Yugoslavia never developed. Instead, Gunn turned his attention to China.

Gunn first visited China in the summer of 1931 on his way back to Paris from New York. His enthusiastic report led Max Mason, president of the Foundation, to send him back in the autumn of 1932 where he remained until February 1934 at which time he returned to present an extensive proposal to the Board of Trustees. The Board went on to approve the program in December 1934. The program was initiated in July 1935.

The first substantive chapter of Gunn's sixty-one-page proposal addressed Medicine and Public health.[8] In it he outlined the need a unified program to be coordinated closely with a program of national planning, since progress would depend upon the "progress in other fields of community activity, such as industry, agriculture, education, and transportation." An administrative arrangement was needed which "correlated" the work of national, provincial, urban and rural units, i.e. the program must derive from national policy and be applied at all levels. Most of the rest of the report dealt with the "other fields of community activity," including the Natural Sciences, the Social Sciences, the Applied Sciences, the Humanities, Primary and Secondary Education, and National Reconstruction. For each field the relevant educational institutions were identified and their potential contribution to this new program evaluated.[9]

The major recommendation that emerged following a review of all these fields was the creation of a program for North China concerned with rural reconstruction that involved the Department of Preventive Medicine of the

PUMC (which John Grant was responsible for); The Mass Education Association at Ting Hsien (Ding County), Hopeh Province; the North China Industrial Institute; the Institute of Economics at Nankai University; the Institute of Rural Administration, Yenching University; and the College of Agriculture, Nanking University.

Gunn briefly indicated how important the Mass Education Association had been for the development of rural public health work. In fact, its history provided the key for his whole proposal. Headed by James Yen, this Association was more popularly known as the Mass Education Movement (MEM).[10] Educated at Yale University, Yen founded a teaching method for the illiterate Chinese population based on the use of around 1000 characters of the alphabet.[11] He first experimented with this approach to literacy in several major cities, beginning with Peking. When he started working in rural areas, "we realized that literacy alone was not enough. Literacy isn't education—it is only a tool for education, a means to the whole end. The people had to get an education which involved the whole of their life."[12]

When the China program was under discussion by the Board of Trustees in 1934, another issues raised was whether "the welfare of mankind (is) best served by enlarging our investment in China? . . . Is there no other sector of the world where we can hope to obtain as large a return in human happiness and welfare as we can in China . . . . We would point . . . to the possibilities that lie in a country like Mexico."[13]

Whether it was with this in mind that led Mason to send Gunn to Mexico several months earlier is not clear. In any case, that visit is of interest as it demonstrated again how free Gunn was in seeking 'investments' outside the health sector. In his *Notes and Comments*, the most important part (6 pages) dealt with the subject of rural education. Here he described the philosophy of the Mexican revolution to make the rural school the "real cultural and social center of village life."[14] After his visit to about 25 rural schools, Gunn reported:

> I am frank to say that seldom have I been so inspired. The great strength of the movement is that the rural school is not something that is imposed on the people by the Government, but it is something that they desire most fervidly and who which the local community has made very material financial and personal sacrifices. One of the most significant activities is that in connection with the development, and even survival, of native industries. The whole scheme involves, of course, an attempt to raise the social and economic level of the people. Furthermore—and this is of vital importance—the emphasis of the whole movement is directed towards a form of education and advancement which will keep the people *in* the country. . . . It is regrettable that, up to the present time, the National Health Service has not cooperated more actively in the federal rural school program. There is a real opportunity to put over a rural health program under the best of auspices and at a minimum cost.
>
> There are points of similarity between the program described above and the activities of James Yen or Ting Hsien in China and also to the work which

> was developed in Yugoslavia by Doctor Štampar in connection with the House of Health in that country. The great advantages in the Mexican situation over China are that it is an activity fostered by the Government, that it can be carried on apparently effectively without many experts, and that it is extremely cheap.[15]

Gunn's report did not lead to any further developments in Mexico. For a variety of reasons, mostly linked with the revolutionary period that Mexico was then living through, he was forced to conclude that "it was not a propitious time to consider enlarging our present small program in Mexico."[16] The program in China was effectively halted when the Japanese attacked China in 1937. Gunn did not get another chance of working on such a broad-based approach to human welfare; he died in 1944 while working on a study of voluntary health organizations in America.

Further excerpts from the introduction that he wrote for the Bandoeng report reflect where his thinking had arrived at that point in his career:

> Preventive and social medicine should permeate the entire programme of medical education. One thing is certain . . . unless the economic and cultural level of the rural populations can be raised, there can be no hope of employing curative or preventive measures with any degree of success. If this problem (land reform) is neglected, programmes of rural reconstruction not only will be greatly retarded, but will not be able to rest on a permanent basis. The opening of public health work in rural areas can often be used as the entering wedge for the development of a broader programme embracing education, economics, sociology, engineering and agriculture.[17]

These words led Halfdan Mahler, WHO Director-General, to note their similarity with the approach of primary health care as developed and promoted under his leadership in the 1970s.[18]

The Evolution of Paul Russell's Approach to Public Health

Paul Russell joined the Rockefeller Foundation in 1923. His first assignments took him to America's South where he participated in various efforts to control malaria. In 1929 he went to the Philippines where he studied the epidemiology of malaria and explored ways for its control until his departure to India in 1935.

Russell concluded that as far as the rural areas in the tropics were concerned, "the problem of malaria control is still unsolved."[19] None of the methods that had been developed with some degree of success in the American South could be afforded. What was needed were "naturalistic methods of species control in the hope of finding cheap automatic procedures, especially those which will lead directly to improved agricultural yields." Whatever methods were to be used, they must be embedded in a long-term strategy:

> Malariologists . . . must visualize inexpensive but continuous programs extending over decades of time. Perfection and speed in any line of human endeavor are expensive. . . . It is probably a mistake for the tropical malariologist to strive for perfection, because the tropics are too poor to pay for it . . . *Time* more than *Money*, and *Continuity* rather than *Perfection*—these must be the mottoes guiding malaria control in the tropics.[20]

Russell toured India for six months in 1935. Many of his first impressions confirmed what he had already concluded from his recent stay in the Philippines. What he called "malaria prophylaxis" was simply not feasible in the rural tropics. Control methods that had worked in America averaged $.70 to $.80 per capita but the communities of the Philippines and India could not afford more than $.05 per annum for control.

Russell marveled at the complexity of the malaria problem in India. Whereas in all of Europe there were ten or twelve anophelines, of which two or three were potentially important, in India there were fifty-three, with no less than thirteen "under suspicion as carriers of malaria." Some bred in "classical marshes and ponds," while others in "running water, in river pools, in seepage, in salt water, in wells and cisterns, and even in hoofprints."[21]

Land reform, which Gunn had cited to be basic importance for health and human welfare, had emerged in India as a major issue following the 1935 Government of India Act. The radical section of the Congress Party insisted on land being made available to all tenants. They also called for reduction of taxes and a food policy that took into account the nutritional needs of all, especially women and children. Wolf Ladejinsky, an expert on land reform, following a visit to India in 1939, concluded that "piecemeal reforms" would not solve the problems of India's rural poor. "Illiteracy, primitive equipment, poor cultivation of the soil, low yields, low standard of living, unsatisfactory landlord-tenant relations, indebtedness, diseases, and poverty" were too intertwined for single-discipline approaches to work.[22] Russell was aware of the miserable plight of the rural poor but confined himself to the observation that India was a "land of financial stress more severe perhaps than anywhere else, save possible in China... The riches of rajahs and the poverty of peasants alike defeat imagination. . . . It seems impossible to unlink the chains which bind farmers so securely."[23]

One approach that had helped to break the vicious credit system of the money lenders was the cooperative movement which was initiated earlier in the century. By 1936 India had a total of 94,000 agricultural cooperatives with over three million members. Although only covering a small percentage of the rural population, it had helped make credit available at more affordable interest rates. Ladejinsky judged that it was to the cooperatives that India should look to increase the number of cultivators willing to use better seeds, improve methods of cultivation, take better care of cattle, adopt precautions against human, animal and plant disease, and market their goods more efficiently.

Whether Russell was aware of the cooperative movement is not clear; he did learn, however, of the more than a thousand cooperative anti-malaria and anti-mosquito societies that had been formed in Bengal but he saw little role for them in malaria control. Like elsewhere in the world, they sprang up in the time of epidemics only to collapse once the immediate dangers are over. Having concluded that "continuity of effort" was an essential feature of any malaria control effort, such volunteer agencies had little role to play.

Russell avoided any discussion of the malaria problem that would force a broadening of his approach to its control. His recommendations concerning agriculture, for example, were confined to administrative and technical points. Thus, where major engineering was involved, as for example irrigation works, he recommended that engineers and health officers involved cooperate and that appropriate legislation be enacted to prevent "malariogenic nuisances." He did not hesitate to criticize many of the agricultural methods used by peasants when these methods exacerbated the malaria problem: leaky pumps, waste irrigation water, and ill-kept drains, not to mention the vast water surfaces involved in rice and sugar-cane cultivation.

Russell promoted work that peasants could undertake themselves in "minor control methods," i.e. naturalistic measures: "[t]he deliberate manipulation by man of one or more of natural factors so as to prevent mosquito-breeding, to destroy larvae, or even to deviate adult mosquitoes from man to animals."[24] But for such measures to have any chance of success they must contribute as well to the overall agricultural yield, otherwise there would not be sufficient incentive for them to be adopted.

Both Ladejinsky and Russell emphasized the importance of education. Here, too, one sees how far apart their vision was. Illiteracy, Ladejinsky wrote, quoting from another author, is a curse that "acts as a leaven to start the vicious circle of debt-disease-and-death."[25] Compulsory schooling was needed for children and mothers should be provided the opportunity to gain literacy to help their children stay in school. Russell limited his focus to health education, in the hope that a community would learn not to put up with malaria if "a majority of its population had a *practical* understanding of what malaria cost them, of why it existed, of what measures were available to control it."[26]

After several years in India, Russell wrote on the "social aspects of malariology," but rather than "enter into an abstract consideration of the social implications of malariology," he chose to consider "one phase of the subject, namely, *social obstacles to malaria control*." In this paper he listed a number of obstacles:

- A fundamental absence of educated and effective public opinion
- A far too limited use of sound administrative principles in public health
- A lack of sufficient numbers of specially trained personnel
- A failure among officials to take cognizance of economic considerations involved in malaria and malaria control

- An inadequate knowledge of methods for applying in a practical way the results of research.[27]

At no point in this article is Russell tempted to engage the "abstract" social aspects of disease control. There is no reference to land reform or the economic situation of the peasants, other than the observation that "malaria frequently maintains poverty." Even his side references to nutrition and housing are made exclusively from the point of view of malaria control. Thus his observation that while "the provision of proper food and shelter is ... an important social health service . . . there is no proof that one can control malaria by increasing the food supply or supplying better-ventilated homes."[28]

Concluding Observations

Selskar Gunn's early career coincided with the period where public health in America was pushed and pulled in many different directions. His early interests reflect the wide-ranging agenda that public health had cut out for itself at the time: volunteer societies in the detection of tuberculosis cases (1910), milk supply (1911), cooperation between health and police departments (1912), housing conditions (1912), and a survey of 45 organizations interested in public health (1915). Only in 1917 did Gunn enter the single-disease control world of the IHD. Paul Russell, 11 years younger than Gunn, had no such exposure to alternative approaches to public health. After obtaining his MD degree from Cornell in 1921 and following a two-year internship, he joined the IHD in 1923 where he was sent to the South to work on the malaria problem.

In the 1910s it was still conceivable that public health would not be a medically dominated profession. William Sedgwick, Gunn's mentor, was a leading proponent of the idea that public health leaders should receive a different training from those of physicians. His so-called "*Y Plan*" of medical education called for a common base of study with two branches, one leading to the degree of Doctor of Medicine, the other to the degree of Doctor of Public Health. But, with the financial backing of the Rockefeller Foundation, American and foreign schools of public health were developed in close affiliation with medical schools thus virtually solidifying medical domination of the field.

Lacking an MD degree proved no obstacle to Gunn's early advancement. It was not uncommon at the time for non-physicians to be appointed health officers. When he joined the Foundation in 1917, the director of the IHD was Wickliffe Rose, an educator. Only when Rose was succeeded by Frederick Russell in 1923 did the leadership of the division come under control of "the medical barons."[29] By then Gunn had carved himself out a niche in the Paris office many thousands of miles away from the main office in New York where Frederick Russell resided. When Gunn was elevated to the status of Vice President in 1927 he reported directly to the President. It was around this time that the Foundation underwent a major revision of its program, partly driven by the desire of some Board members to curtail the power of the medical barons. It was during that period that Gunn began to envision a broader role for the foundation in human welfare. John Grant in his oral history described this period as follows:

> Being imaginative, Gunn was intrigued by the problem that in no country was it (the Foundation) able to have a unified multi-purpose Foundation program. This concerned him to a very great extent, and he was constantly bringing it up in conversation, that if such a program could be brought into existence, that the Foundation's goal of the welfare of mankind could be more effectively demonstrated than through each discipline undertaking its own project.[30]

In Grant's opinion, Gunn was a more imaginative public health leader due to his not having a medical degree. He, along with others such as Charles Winslow, who was another one of "Sedgwick's boys" and a close friend of Gunn, "were not chained down by the conventional medical ideas on public health."[31] Under the leadership of Max Mason, who became president of the Foundation in 1929, Gunn again found himself in a situation where he was given free reign to think along unconventional lines.

Mason retired in 1936, i.e. one year before the Bandoeng meeting. His successor, Raymond Fosdick, had been responsible for guiding the changes that Mason, in principle, had been hired to implement. Although these changes were designed in part to curtail the powers of the medical barons, by 1937 the IHD was stronger than ever – "the medical barons were still firmly in the saddle."[32] Paul Russell's approach to malaria was in full keeping with the IHD approach to public health; his IHD colleagues were engaged in similar activities around the world with respect to malaria and a host of other leading infectious diseases.

There is little evidence to suggest that Fosdick had any sense of the importance of Gunn's China program for public health. In his history of the Rockefeller Foundation, he presented it under agriculture, indicating that it had emerged from a concern with plant breeding. He seems not to have been aware that some Board Members (in 1934) had recognized Gunn's program as an approach to public health that was an alternative to the one being promoted by the IHD.[33] Some in the public health community, however, understood its importance. In Gunn's obituary (written by Winslow), one finds:

> The plan which he developed for a broad program of social welfare—in which the artificial boundaries between health and education and agriculture and economics were broken down—was perhaps the most significant achievement of his life. When the menace of Japanese imperialism is overthrown, this program will certainly be realized as a basis for the great emergent China of the future.[34]

On the other hand, the China program was seen as a "tangible expression" of the kind of program he thought the Foundation should be promoting.[35] Unfortunately, both Frederick Russell and his successor, Wilbur Sawyer, did not like Gunn's China program. Russell thought that it was Gunn's eloquence that had allowed him to sell it to the Trustees; Sawyer did not appreciate how loose the whole setup was, i.e. its day to day operations were not under the control of Rockefeller officers. Without the active participation of the IHD there was no chance of the Foundation developing a "broad program of social welfare," as hoped for by Winslow.

While there were malariologists who recognized the importance of a social approach to malaria, when the powers of DDT were discovered in the early 1940s the rug was essentially pulled out from under them. With Paul Russell leading the World Health Organization's Malaria Expert Committee and Fred Soper, another Foundation staff member, operating out of the position of Director of the Pan-American Health Organization, the world's first global disease eradication campaign was launched.[36]

As I've argued elsewhere, while the China program in theory could have served as an example for the World Health Organization (WHO) to follow, in practice, this had little chance of happening.[37] Whereas the Rockefeller Foundation had the potential of promoting a multi-sectoral approach to human welfare, which, with the exception of Gunn's program in China it chose not to pursue, WHO's mandate is far less inclusive. To operate as the Foundation did in China would have required WHO to create some mechanism that brought together the other major UN specialized organization, in particular the Food and Agriculture Organization (FAO), to work together for rural development or the equivalent. Thus, even if DDT had not been discovered, it is hard to imagine how a multi-discipline approach could have taken root in the post WWII world where each discipline (health, education, agriculture, economics) had its own international organization to offer guidance and advice. Other factors further complicated the situation; for example, the Cold War alone generated sufficient obstacles to block any approach to human welfare that smacked of being socialism.[38] At this historical moment the IHD's single-disease oriented approach to public health had everything going for it—practicality, avoidance of destabilizing reforms, and tangible results; that of Gunn was a dead letter.

Notes

Those references introduced by RAC are located at the Rockefeller Archives Center in Sleepy Hollow, NY. Each reference is followed by an indication of the Record Group (RG) number, Series (S) number, Box (B) number and Folder (F) number where it is located.

1. *League of Nations, Report of the Intergovernmental Conference on Far-Eastern Countries on Rural Hygiene*, Bandoeng, August 3rd to 13th (Geneva: League of Nations, 1937).

2. Quoted in Elizabeth Fee, *Disease & Discovery: A History of the Johns Hopkins School of Hygiene and Public Health 1916-1939* (Baltimore: The Johns Hopkins University Press, 1987), 66.

3. *Ibid.*

4. John Farley, *To Cast Out Disease: A History of the International Health Division of the Rockefeller Foundation (1913-1951)* (Oxford: Oxford University Press, 2004), 48.

5. RAC RG 12.1 Gunn Diaries 1928.

6. A Brief Summary of the Conference of Trustees and Officers at Princeton, October, 1930, RAC RG1 S900 B22 F166. appendix 2, p.1.

7. *Ibid.*, appendix 2, p. 4.

8. Much of the first part of the proposed program was developed with the aid of John B. Grant, a foundation staff member who had been in China from 1922 where he was responsible for the development of the public health. curriculum in the Peking University Medical College (PUMC).

9. Selskar Gunn, China and The Rockefeller Foundation, 23 January 1934, RAC RG1 S601 B12 F129 p. 1 attachment to Mason.

10. John Grant subsequently noted that "without Jimmy Yen there would never have been the opportunity for Selskar M. Gunn and The Rockefeller Foundation to have a North China project." *Reminiscences of Dr. John B. Grant*, RAC S900 Hist. Grant, Vols. 1-7, p. 325.

11. The textbook used for adults was entitled: The One Thousand Characters. C.C. Chen (PUMC) helped revise it to include 8 complete lessons and four references to health in the four-month course. Courses were normally organized in the winter when agricultural work was less demanding.

12. Pearl S. Buck, *Tell the People: Talks with James Yen About the Mass Education Movement*, (New York: The John Day Company, 1945), 22.

13. 1934 report, (n. 9), p. 108.

14. Gunn, Notes and Comments on a Visit to Mexico September 2—November 15, 1934. RAC. RG2. S.1938 B.166 F.1208.

15. *Ibid.*

16. Gunn to Mason, November 10, 1934 RAC. RG2. S.1938 B.166 F.1208.

17. Bandoeng report (n. 1), p. 22-36

18. Mahler, H. Promotion of Primary Health Care in Member Countries of WHO. *Public Health Reports*, 1978, 93 (2): 107-112.

19. Paul Russell, "Malaria in the Philippines," *American Journal of Public Health*, 1936, 26 (1): 5.

20. *Ibid.*

21. Paul Russell, "Malaria in India: Impressions From a Tour," *American Journal of Tropical Medicine* 16:6 (1936): 654.

22. Wolf Ladejinsky, *The Selected Papers of Wolf Ladejinsky: Agrarian Reform as Unfinished Business* (Oxford: Oxford University Press, 1977), 38-9.

23. Russell "Malaria in the Philippines," 654.

24. *League of Nations, Report of the Intergovernmental Conference,* 92-3.

25. Ladejinsky, *The Selected Papers of Wolf Ladejinsky*, 37.

26. Paul Russell, "Some Social Obstacles to Malaria Control," *The Indian Medical Gazette*, (1941): 683.

27. *Ibid.*, 682.

28. *Ibid.*, 688.

29. Discussed in Farley, *To Cast Out Disease*, 158-165.

30. *Reminiscences of Dr. John B. Grant*, RAC S900 Hist. Grant, Vols. 1-7, 449-50.

31. *Ibid.*, 449.

32. Farley, *To Cast Out Disease*, 165.

33. See Socrates Litsios, "Selskar Gunn and China: The Rockefeller Foundation's 'Other' Approach to Public Health," *Bulletin of the History of Medicine* 79 (2005): 295-318.

34. Editorial, *American Journal of Public Health*, 34:10 (1944): 1096-7.

35. J.E. Harr, and P.J. Johnson, *The Rockefeller Century* (New York: Charles Scribner's Sons, 1988), 365.

36. See Socrates Litsios, *The Tomorrow of Malaria* (Wellington, New Zealand: Pacific Press, 1996).

37. See Litsios, "Selskar Gunn and China."

38. See Socrates Litsios, "Malaria Control, the Cold War, and the Postwar Reorganization of International Assistance," *Medical Anthropology* 17:3 (1997): 255-278.

Chapter Five

Globalizing Primary Care: Rockefeller Philanthropy and the Development of Community-Based Approach to Public Health in Sri Lanka—What Can We Learn?

Soma Hewa

The Alma-Ata Declaration of the World Health Organization (WHO) in 1978 demanded immediate action by all governments and international organizations to promote the health of all people around the world. The Declaration has become the guiding thread for action on health promotion in both developed and developing countries. The main argument of this declaration has been reenforced by subsequent conferences held in Ottawa (1986), Adelaide (1988) and Jakarta (1997). While the Alma-Ata Declaration reaffirms the original *Charter* of the WHO, which recognizes "health as a state of complete physical, mental and social wellbeing," it recommends that all nations must provide "essential health care based on practical, scientifically sound and socially acceptable methods and technology, made universally accessible to individuals and families in the community through their full participation and at a cost that the community and country can afford to maintain."[1] The Declaration makes explicit reference to "equality," "social justice," "community participation," and most importantly to "overall social and economic development of the community" as fundamental components of the primary health care system.

The Alma-Ata Declaration on primary care represents a marked shift in the approach to health from curative medicine to building equity in social and economic well-being. Given the emphasis on the overall wellbeing of the community, it is no surprise that this declaration has become a landmark document in health promotion and international co-operation in recent years. Further, it is important to recognize that the Alma-Ata Declaration was made on the eve of a new global era that began with the end of the Cold War. It argued that the new international economic order is the basis of health for all people, and sustained economic growth is dependent upon the protection of health.

Although this new model of primary health care proposed by the Alma-Ata Declaration seems to represent a novel concept, basic models of universally accessible, community-based health programs were developed in a number of developing countries as early as the beginning of 20th century.[2] One such model was the Health Unit program of the International Health Board (IHB) of the Rockefeller Foundation, which helped to create a primary health care system in Sri Lanka. This program established a community-based public health network across the country, which became the basis of the national health policies during the last fifty years. The outstanding health status in contemporary Sri Lanka is the result of these early efforts to build a comprehensive network of primary care services throughout the country.

There is an important lesson to be learned from this program for both developed and developing countries. The community-based primary care programs may be the answer to basic health problems in developing countries, where communities are well integrated by traditional social values. A primary care program designed to incorporate these values and local resources, backed by technical support from the international organizations, could bring lasting results in promoting health. In developed industrialized countries, advanced medical services are widely available in major urban centers. However, in rural and aboriginal communities basic facilities such as running water, proper sewerage systems and adequate housing are still major problems. Unless these problems are addressed by social and economic programs, diseases such as tuberculosis and meningitis, which are major health problems in these communities, could easily spread to rest of the society. Further, social problems such as alcohol and substance abuse, suicide and violence could undermine social and political stability in these countries. The basic strategies of the health unit program developed by the IHB provided a long-term vision, which recognized the importance of international co-operation in combating disease and promoting health. This program influenced the global health agenda of the WHO.

Historical Background

International co-operation in health began in a campaign to prevent the spread of disease rather than to promote health. Like most national health programs, international health was born out of the fear of disease. It was the threat of cholera in the early 19th century that brought nations together. Asiatic cholera, which traveled through the newly established trade routes between Europe and its colonies in Asia, Africa and North America, devastated major urban centers in Europe. As one historian put it, "Terror, panic, fear, alarm, fright and dread were all words used to describe people's reactions to the prospect and impact of cholera in 1832."[3] What followed was a series of international conferences to develop global strategies to prevent the spread of the disease rather than to build a broad front to promote health. In 1851, the first international sanitary conference held in Paris recommended strict quarantine measures for travelers and merchant sailors suspected of carrying diseases.

Following this conference, a coordinated campaign was commenced to prevent the spread of cholera, plague and fever from Asia to Europe and to North America.

Although by the end of the nineteenth century, scientists were able to identify the specific pathogenic organisms responsible for diseases such as cholera, tuberculosis, typhoid and plague, the real effect of these discoveries began only decades later, when antibiotics became widely available. In the meantime, however, the campaign to create a permanent body to monitor the outbreak of infectious disease and to develop international policies to control the spread of disease resulted in the founding of *l'Office International d'Hygiene Publique* (OIHP) in 1909. The mandate of this organization was to collect epidemic intelligence on specific diseases on behalf of member states: "To collect and bring to the knowledge of the participating states the facts and documents of a general character which relate to public health and especially as regards infectious disease, notably cholera, plague, and yellow fever, as well as the measures to combat these diseases."[4] Although international co-operation to prevent the spread of disease improved with the creation of the League of Nations, the emphasis was still on the control of disease. The Covenant of the League of Nations, article 23 (f) stated that the objective was to "endeavour to take steps in matters of international concerns for the prevention and control of disease."[5] In view of this particular objective, the Geneva Office of the Health Organization of the League of Nations gathered and analyzed data on the geographical origin and the patterns of distribution of major epidemic diseases such as cholera, plague, yellow fever, dysentery, typhus and many others that generally originated in the European colonies and occasionally traveled to the West.

In the early 1930s, the *expert committees* of the Geneva Office recognized the importance of nutrition, housing, medical education and sanitation services for health. The work of these *expert committees* had far-reaching effects on a wide range of public health issues, including minimum standards of diet and nutrition throughout the world. Although several international conferences organized by the Geneva Office, between 1931 and 1937, identified public health needs in developing countries, large scale international health campaigns during this period remained the responsibility of major philanthropic foundations and voluntary organizations. Among the most notable was the International Health Board (IHB) of the Rockefeller Foundation, which was the first to undertake mass treatment campaigns against hookworm disease and yellow fever around the world. Victor Heiser, the director of the Asia-Pacific program of the IHB in the 1930s, summed up the specific concerns that inspired the IHB in these words: "I had not overlooked the fact, repeatedly impressed upon me in my immigration training, that disease never stays at home in its natural breeding places of filth, but is ever and again breaking into the precincts of its more cleanly neighbors."[6] Although the concerns about the threat of disease influenced the decision to undertake mass treatment programs in

European colonies, the IHB was willing to accept new challenges and move beyond the standard practices in international health during that period. Unlike other international organizations involved in health, the IHB commanded the most resources and the expertise to initiate large-scale public health campaigns around the world.

Health Unit Program in South Asia

Following the successful hookworm treatment program in the southern United States during the early 1900s, the IHB extended its medical expertise to other countries, in particular the British colonies. The IHB began a mass treatment campaign for hookworm disease among Indian migrant workers on the plantations in Sri Lanka in 1915. The campaign received lukewarm support from the British planters and the colonial government in Colombo. The IHB representatives had to walk a fine line between the colonial government and the powerful British planters, who constantly locked horns over the responsibility of providing public health services. Although the campaign failed to control the disease, it succeeded in reducing the intensity of infection among the migrant workers. After nearly a decade of hookworm treatment programs on plantations, villages and towns, the IHB representatives in Sri Lanka had a better understanding of the magnitude of the country's health problems, public health requirements and social and economic conditions. They became convinced that the piecemeal treatment programs for hookworm infection would not bring long-lasting results without addressing the cause of the wider range of health problems. The solution, according to the IHB, was to develop an integrated approach in which hookworm control was part of a larger public health campaign. Consequently, the Health Unit program was developed.[7]

Although the IHB had experimented with the health unit system in the southern United States, it did not transplant that particular program in Asia. While recognizing the universality of public health principles, W. P. Jacocks, the IHB representative in Sri Lanka, emphasized the fundamental importance of concentrating on the specific health requirements, social and economic conditions of each community in which health units were established: "The Health Unit has been developed to meet the health needs of populations living in rural and semi-rural areas. It is based on standard public health organizations in which all recognized health activities are carried out." The key strategy was to identify communities that represented a unique set of problems and address their root cause in order of their importance for morbidity and mortality. "In undertaking general public health measures," argued Jacocks, "it is considered advisable, by those most experienced, to begin work on a small scale in a restricted area. A lively campaign against *all* preventable diseases should not be immediately undertaken. Even were this possible, it would be unwise, for the method of attack may vary in different countries. The preventable diseases peculiar to the district should be determined by careful survey and they should be thoroughly studied and attacked in the order of their importance as a cause of morbidity and mortality. The work should then be gradually enlarged to include all the public health problems in the area."[8]

Organizational Structure of the Health Unit Program

Once the health unit was established, the first step was to conduct a survey to determine the scale and the nature of health problems in the community. According to Jacocks, "Know your area, and know your people is the slogan of Ceylon Health Unit work." It was a pre-requisite for all the members of the health unit to become fully acquainted with the people, as well as the socio-economic and environmental conditions of the community.[9] Based on the information gathered by the survey, a program was prepared in consultation with the representatives of local and central governments. Each health unit, serving up to 80,000 people, undertook the standard work of a public health department such as the development of sanitary programs, health education, vaccination of children, maternal and infant welfare, school hygiene program and the collection and study of vital statistics.[10] In view of these specific tasks, each health unit recruited one medical officer (primary-care physician), five public health nurses, ten mid-wives, five sanitary inspectors and a secretary responsible for maintaining records such as vital statistics of the health unit area. These full-time public employees were the core staff of health units. In addition, a dentist, an eye specialist, a nose, throat and ear specialist, and an orthopedic surgeon from the district hospital visited health unit clinics periodically [See Figure 5.1].

Responsibilities of Personnel

The physician, who also served as the director of the health unit, coordinated all activities to ensure that "public health inspectors, nurses, and midwives carry out the work as planned. This is accomplished by making frequent visits to each member of the field staff in every part of the area."[11] He/she investigated health problems in the area, made visits to schools, conducted clinics for vaccinations, maternity and child welfare services and provided health education through public lectures and propaganda. The health units provided limited curative services and when necessary referred patients to the local hospital. Further, the physician was also responsible for bringing social and economic problems of the community that had health implications to the attention of local authorities and government departments for action. For example, the problem of nutrition among the rural population in Sri Lanka was one of the major concerns of the Health Unit program. Accordingly, the program recommended that health units must give priority to improve the economic condition of the peasant, which was "a matter for the coordinated action of several departments."[12] In collaboration with other government departments, the health units introduced supplementary nutritional programs for expectant and nursing mothers and infants. This included the distribution of milk and milk powder, cod-liver oil and vitamin tablets free of charge. In addition, nutritious and well-balanced midday meals were provided for schoolchildren in rural districts. Regular surveys were conducted to determine the nutritional requirements and to evaluate the progress of on-going activities. As a part of the health education campaigns, the health units developed a nutrition guide advising on how to prepare affordable and healthy daily meals using the available fruits and vegetables grown in most villages.

The public health nurses and midwives were directly responsible for maternal and child welfare activities. It was in this category of work that the Health Unit system provided the greatest service in Sri Lanka. In 1932, describing the valuable services health units rendered to the promotion of maternal and child welfare, the deputy director of the Department of Medical and Sanitary Services argued, "The most popular branches of Health Unit work were the Maternity and Child Welfare sections. The organization of the Health Unit lent itself with the greatest facility to extension to the rural areas."[13] The public health nurses and midwives organized routine clinics and visited schools to conduct health education programs, and to administer vaccinations, hookworm and malaria treatments. As the public health nurses and midwives were directly involved with the community, they were responsible for enlisting all expectant mothers and children in the community for monthly clinics. In a report prepared for the IHB requesting assistance to establish a public health nursing program, Jacocks argued that if "the Health Unit fails to succeed and develop in Ceylon, it will be for lack of public health nurses."[14]

The midwives were responsible for locating expectant mothers, arranging early medical examinations, and attending the birth if the pregnancy was normal. Following the birth, they were required to visit the mother and newborn daily for ten days. In case of medical complications during pregnancy, the midwife ensured that a doctor was called or that the mother was taken to the hospital. Given the importance of maternal and child welfare activities for health units, the IHB continued to provide expert advice on training personnel for this work. A standard set of guidelines was developed for all midwives regarding prenatal to postnatal care to ensure the health of both mother and child.

Another important aspect of the Health Unit program was the extensive sanitary projects undertaken in both urban and rural areas. As basic sanitation services were almost non-existent in most parts of the country, the IHB provided new ideas and methods to create adequate and sustainable public sanitary programs in the health unit areas. In this connection, health units in predominantly rural areas provided functional, low cost latrines and rural wells. The annual report of the Health Unit program for 1933 provides a detailed account of the number of latrines and wells constructed during that year. According to the report, a total of 2,424 household latrines were constructed in all health unit areas. In addition, 563 latrines were constructed in schools. These numbers represented a significant increase from the previous years. The IHB took particular interest in providing adequate latrines as a public health measure. Commenting on the success of the sanitary program in health unit areas and recommending further support from the IHB, Jacocks wrote, "Ceylon has been foremost in developing rural health work and is going forward steadily."[15] Along with continued treatments for hookworm infection, the extensive public sanitation services such as the construction of latrines, sewerage systems, the malaria control campaigns and better water supplies across the country greatly improved the health of the population.

Community Support

The basic philosophy of the Health Unit program was to build a network of health services through the co-operation of the people in all aspects of the work, so that the community was responsible for providing resources and leadership and, above all, carrying out the activities assigned by public health officials. Such cooperation would ensure the sustainability of the program. The international experts were to provide technical support and arrange additional resources as deemed necessary. In a report titled *A Guide to Health Unit Procedure,* Jacocks and his Sri Lankan counterpart S. F. Chellappah stated, "To get lasting results the work must be placed on a co-operative basis which is the foundation of the Health Unit system. Co-operation is obtained by carefully explaining . . . the objectives to be attained and assistance which those concerned may give in helping to carry out the program."[16] Clearly, the Health Unit program was designed to draw a collective response to local health needs and to utilize local resources and people. The three levels of co-operation—individual citizens, community organizations and local authorities—were sought and received by the Health Unit program. Ten years after the first health unit was established, the authors of the program expressed satisfaction with the support it had received at various levels.

At the individual level, private citizens donated buildings to set up dispensaries and clinics: "Public spirited people have rented and furnished buildings for holding weekly clinics. Other people have donated lands, buildings and furniture for maternity homes; some have provided pumps for wells, and a few have offered 'Challenge Shields' to stimulate Health Education."[17] It is no doubt that the Health Unit program relied heavily on the generosity of the people who believed that it was their responsibility to be part of the program. Likewise, at the community level, social service organizations were created to assist public health workers involved in child welfare programs such as immunization and the distribution of milk. In 1931, there were twenty-four voluntary organizations under such names as "Social Services Leagues," "Health Leagues" and "Child Welfare Leagues" working with health units. During that year, they had contributed Rs. 16,802 to health units across the country.[18] In addition, Social Services Leagues provided free sewing lessons for young mothers. These voluntary organizations were "supported by donations, monthly contributions and special collections, and their activities [were] not a charge against the Health Unit budget."[19] The local authorities, such as municipal and urban councils, welcomed the program as it enabled them to incorporate their regular health and sanitary work with the health units without having to bear additional financial burdens.

Cost of Health Units

According to the available data, the total cost of the Health Unit program amounted to a small fraction of the annual health care spending in Sri Lanka. By 1933, a total of 8 health units had been established across the country. Although the cost of operating health units varied depending upon the size of the population in each health unit area and the magnitude of the public health problems, the average annual cost of a health unit in 1933 was Rs. 240,205 (approximately $81,000). This represented about 2.6 percent of the total annual budget of the Department of Medical and Sanitary Services. The per capita cost of the Health Unit program was about Rs. 1.03 (approximately $ 0.34). It should be noted that the IHB did not assume direct financial responsibility to the Health Unit program. Its financial support was limited to what it described as "special circumstances" such as an unexpected shortfall of government funding for established programs or the need for recruiting additional personnel for specific projects sponsored by the IHB.[20] When considering the amount of public health work carried out by the health units in Sri Lanka, the cost of the program was negligible.

One of the most important "special projects" that the IHB financed was the training of public health workers for Sri Lanka and other South and Southeast Asian countries, where the IHB had begun public health activities. The first health unit in Asia was established at Kalutara, Sri Lanka as a pilot project in 1926. It became the model for all the other health units in the region. In 1928, the IHB expanded the Kalutara health unit to provide training facilities for public health personnel, such as public health nurses, midwives and public health inspectors for health units in other parts of Sri Lanka as well as India, Burma and Malaya. In recommending a $4,440 grant for the Kalutara health unit, Jacocks wrote: "Experience has confirmed the wisdom of having the first health unit in the East in Ceylon. Owing to its central location, Ceylon has served as a convenient training ground and demonstration center for health officers from Burma, Madras, [and other Indian cities], federal Malaya State, Java and Sumatra, and visitors from many countries have seen the work."[21] These foreign public health workers received two weeks to six months training on various aspects of the prevention of disease at the expense of the IHB. Arguing in favor of expanding the facilities at Kalutara, Jacocks maintained, "the Kalutara [health unit] has been, and continues to be, the chief training center for the East. Without the facilities, which it offers our rural work would be immeasurably handicapped. . . . If plans in India materialize there will be four or five officers sent to Ceylon for training in 1936."[22] The importance of the Kalutara health unit continued to grow as a training center of public health personnel in Sri Lanka. With the inception of malaria control work in the late 1930s, a large number of field medical officers were trained at Kalutara. As the demand for public health personnel increased in Sri Lanka during the post-independent period, the Kalutara health unit became the National Institute of Health Sciences, which today is the premier training center of public health personnel in the country.[23]

Long-term Results

The Health Unit program must be understood against Sri Lanka's long-term public health achievements. With the establishment of health units in 1926, curative and preventive services were brought under one administration. In 1931, the Departments of Local Government and Health were further unified under one Ministry by giving greater responsibility to local administrations for public health matters. All these administrative changes, which were necessitated by the Health Unit program, laid a solid foundation for the development of preventive and curative health services in the country. Of particular importance were the vaccination of school children and the eradication of malaria, which together contributed to increase life expectancy. Several years later, the International Bank of Reconstruction and Development (later World Bank) observed that the IHB's involvement in public health in Sri Lanka was "an important landmark in the development of public health services of the country."[24] These programs provided the much-needed basic services for the rural population that later became the basis of the government-sponsored primary health care system of the country.[25] Before the establishment of rural clinics by the health units, public health was the responsibility of the assistant sanitary officers of the district. Their work was mostly confined to the urban centers while the rural population was largely ignored.[26] From the outset of the Health Unit program, considerable attention was given to expanding health services into rural areas, with the result that the number of health facilities increased rapidly. One important outcome of this expansion was that a corresponding emphasis soon came to be directed toward training public health personnel to staff these facilities. In 1932, the IHB sponsored a survey to determine the conditions of public health nursing in Sri Lanka. The results of the survey recommended that if the health units were to be expanded in the country, it was of vital importance to establish a training program to ensure "a growing supply of public health nurses so that standardized personnel may be available as the new units develop."[27] The potential of the rural clinics was further realized when they were later expanded as rural "dispensaries," which provide the primary care throughout the country.

Since political independence in 1948, Sri Lanka has established one of the most extensive public health care systems in the developing world.[28] Today, every Sri Lankan has access to a primary health care hospital within a radius of two miles. Unlike in most other developing countries, the problem of access to basic health care "virtually does not exist in Sri Lanka."[29] Approximately 95 percent of births are now taking place in a medical institution, and more than 97 percent of children are fully immunized against major communicable diseases. Although it ranks among the world's low-income countries, Sri Lanka's record in public health has often been compared to the records of industrialized Western countries. High life expectancy (seventy-two years) and low mortality rate (five per 1,000) on the island approach the level of high-income countries.[30] These achievements, no doubt, are the result of a host of public health and social programs, which have been introduced during the last seventy-five years.

Implications for Global Health

What specific lessons can we learn from the fundamental principles and methods of the Health Unit program, both as a project sponsored by an international organization and as a community-based public health program aimed at preventing disease and promoting health? In the following section, we address the implications of the health unit program for shaping the contemporary global agenda for health.

The Health Unit program evolved from and responded to the growing concerns in the 19th century about the threat of disease spreading across international borders. This concern was born in a particular historical context in which expanding global trade and mass migrations created ideal conditions for the exchange of diseases. This was the first stage of globalization in the modern period. The initial reaction was to minimize the export and import of disease through international regulations, which also undermined the free flow of goods and labor. As the volume of trade across territories increased, the threat of disease required new strategies that would reduce the prevalence of disease. The developed countries and international organizations responded by extending scientific knowledge and methods to combat disease. These too became ineffective as the recurring epidemics of diseases in economically impoverished countries found their way to the more affluent neighbors in the West. Thus, international health policies designed to control disease were gradually expanded to include population health promotion strategies in local communities. The Health Unit program was an important step in that direction. The cholera epidemic and plague in the early 19th century, the influenza pandemic, HIV/AIDS and the West Nile virus in 20th century began as local problems and soon became global problems because the world is open for business. If disease has no geographical boundaries in an open world, no country or region is protected from infectious diseases. Thus, all nations are interdependent for achieving health. Health is a global public good in that "the best way to ensure one's own well-being is to be concerned about that of others."[31] There is almost universal consensus today that 1) Health is a key concern in economic globalization, 2) There is a definite benefit to all nations by improved health in developing countries, "the Global Public Good," 3) Health gains/benefits are much greater when specific policies are developed to address "population health," rather than individual risk factors, 4) Health gains are much greater when nation states, international organizations and civil society groups can converge as alliances to address national and international health concerns.

The current phase of economic globalization has intensified the means of spreading infectious diseases across the globe due to rapid and improved transportation networks. The recent wave of SARS is a case in point that began as a public health problem in a remote agricultural village in China, traveled across Southeast Asia and then to North America. The disease caused severe financial burdens to the health sector in the countries affected, in addition to the economic losses associated with the decline of tourism and the hospitality industry. Taking this as an example to highlight both epidemiological and economic implications, the proponents of global public good theory argue that global control of infections is clearly an item to be included, along with international security and peace, within the agenda of the United Nations for action. They argue that so long as rampant disease persists in some parts of the world, it poses a serious threat to economic development and political stability everywhere else. Tuberculosis, malaria and HIV/AIDS have had a devastating impact on already impoverished countries in central and southern Africa. Between 1965 and 1990, countries with high malaria incidents recorded only 0.4% annual growth rate of per capita income, compared to that of 2.3% in non-malarial countries. Countries with intensive malaria epidemics had a negative growth during the same period. If this trend continues, the disease will have definite epidemiological and economic impact on the rest of the world.[32]

The proponents of the global public good thesis suggest that the cost of providing basic care in poor countries should be a part of the overall health care budget of developed countries. This has gained some recognition in the United States as several federal departments include spending for overseas health programs as part of their budget. For example, the Center for Disease Control (CDC) justifies budgetary allocations for global health activities on the grounds that global health is a matter of national security. Likewise, the United States Departments of Agriculture, Labor, and Defense provide funding for the global fight against HIV/AIDS.[33] This spending is not a part of the overseas development assistance. It has been estimated that the effective control of HIV/AIDS, tuberculosis and malaria in Africa alone would require an annual budget of $7-10 billion. Globalization brings many competing economic, social and political forces together. Therefore, it is important to create broad alliances to fight disease and promote health globally as all nations have a stake in it. As seen in the Health Unit program, any plan that is designed to address health concerns in communities must have genuine cooperation among national governments, international organizations that promote specific economic and trade policies, multinational corporations, and various civil society groups. A number of recent high profile broad-based policy initiatives, such as the Global Alliance for Vaccines and Immunization (GAVI), Global Fund for the Fight of AIDS, Tuberculosis and Malaria, Medicine for Malaria Venture, and Stop Smoking

Campaigns, by the United Nations, WHO and nonprofit organizations have given impetus for the creation of international alliance.[34]

Lessons for Developed Countries

What specific lessons does the Health Unit program offer for developed industrialized countries such as the United States and Canada, which are at the forefront in the development of scientific medicine and medical technology as part of their national health policies? What is the state of primary care in these countries?

In the developed industrialized West, there is a growing disenchantment with the declining return in improved health and reduced morbidity for an ever-increasing national health care spending. In the United States, annual health care spending is about 14 percent of the GDP. Yet, some Native and rural African American communities and inner city low-income districts have the extremely poor health status characteristics of developing countries. For example, studies have shown that mortality rates in Harlem, one of the most deprived areas in New York City, are significantly higher than those in rural Bangladesh for most age groups.[35] In Canada's rural and northern aboriginal communities, tuberculosis, meningitis, alcohol and substance abuse are chronic problems. The mortality rates among aboriginal people are several times higher than those of the rest of the population. Situations like these have led North Americans to question the rationale of continuous spending on medicine and medical technologies.

This debate has intensified in recent years by the resurgence of old infectious diseases such as typhoid, tuberculosis, hepatitis and meningitis, which were thought to have been conquered in the early 1970s. The resurgence of these types of diseases certainly suggests that the prevailing understanding of the power of the "magic bullet" or antibiotics is a myth. While most of those old diseases began to decline in the affluent West because of improved living conditions long before the development of scientific medicine and vaccinations, the resurgence of these diseases is clearly due to the breakdown of infrastructure and growing poverty in specific communities in the West.[36] In 1992, the National Institute of Medicine in the United States identified six major factors that contributed to the emergence of disease: 1) The breakdown of public health measures, 2) Economic activities and land use, 3) International travel and trade, 4) Technology and industry, 5) Human demographics and behavior, 6) Microbial adaptation and change. The resurgence of previously eliminated infections in developed countries is mainly due to the neglect of public health services particularly amongst the poor and inner city populations. For example, in the early 1990s, the outbreak of tuberculosis in major cities in North America sent a powerful warning that the failure to provide adequate housing for the poor and homeless created an ideal breeding ground for the disease. The failure to maintain vaccination programs for measles and whooping cough resulted in the return of the diseases in epidemic proportions in the early 1990s. An equally important factor in the resurgence of infectious disease in North America is the failure to upgrade the infrastructure, such as sewerage and water supply systems, in the face of expanding cities and growing populations. Although diseases like typhoid and cholera remain rare in the United States, "various other water and food-borne diseases . . . kill thousands each year."[37] In 1993, for example, 400,000 people became sick in Milwaukee because their drinking water had become contaminated with waste from a nearby dairy farm.

Likewise, there is a crisis in the primary care system across Canada. In major urban centers, rural and Northern territories, the shortage of family practitioners often leads to overcrowding in hospital emergency rooms causing strain on already stretched resources in the health care system. This is a direct result of the reduction of general practitioners graduating from medical schools in the 1990s. The Health Manpower Committee, created in 1982, recommended that the number of general practitioners be reduced by encouraging trainees to pursue certain specialties, such as surgical and laboratory procedures. The committee recommended a further reduction of postgraduate training of general practitioners by 125 per year.[38] This was the period that the expansion of hospitals equipped with high-tech medicine became the priority of the Canadian health policies.[39] The overall neglect of primary care is evident in Canada's aboriginal communities. In Canada's north (Northwest Territories), the physician-population ratio is 1:1068 and the average distance to the nearest physician is more than 150 kilometers.[40] The social and economic deprivations in the aboriginal population are reflected in high morbidity and mortality rates. Poverty, unemployment, overcrowding and sub-standard housing have led to suicide, alcohol and substance abuse and chronic and acute respiratory diseases, which take a heavy toll among aboriginal people. The mortality rate among the aboriginal population is more than double the rate for the general Canadian population, 15.9 compared to 6.6 deaths per 1,000 population, and has been the pattern for more than a half century.[41]

The breakdown of public health infrastructure in rural communities and small towns due to spending cuts in recent years has resulted in an increase of water-borne diseases. For example, in Walkerton, Ontario, seven people died and thousands more became ill in 2000 when their town's poorly disinfected water system became contaminated with e-coli bacteria from a nearby farm runoff. As expanding commercial farming and human settlements move closer to each other, chemical wastes are getting into ground and surface water contaminating drinking water sources. In Canada, tests on private wells show that 40 percent routinely confirm levels of contaminants exceeding standards of potability.[42] The public health infrastructure across Canada is in poor shape and requires major capital investment to fix. Yet, overall, health care spending has increased steadily for the last twenty-five years. In 2005, Canada's health care spending was approximately 10.5 percent of the gross domestic product (GDP), compared to that of 9.1 percent in 2000. A large proportion of that, nearly 47 percent, has gone to hospitals and drugs.[43]

The disparities in morbidity and mortality rates among different social groups in the United States and Canada are due to social and economic inequalities. Several studies have shown that the key socio-economic determinants of health are poverty, unemployment, level of education and social environment.[44] There is no question that these countries have the most sophisticated and expensive medical services anywhere in the world, yet social and economic deprivations remain the major obstacle to better health for these communities. Unlike large-scale capital spending in advanced medical technologies, investment in community-based health promotion activities, as seen in the Health Unit program, lead to greater returns in improved health. By investing small amounts of capital to improve basic public health requirements at the community level, the Health Unit program enabled Sri Lanka to make substantial health gains in the long run. In contrast, despite the significant increase in health care spending in North America, life expectancy has not increased correspondingly in the past three decades because of social and economic inequalities.[45] Therefore, what is now required is to address social and economic disparities, which are responsible for high mortality rates in North America, through community-based development programs.

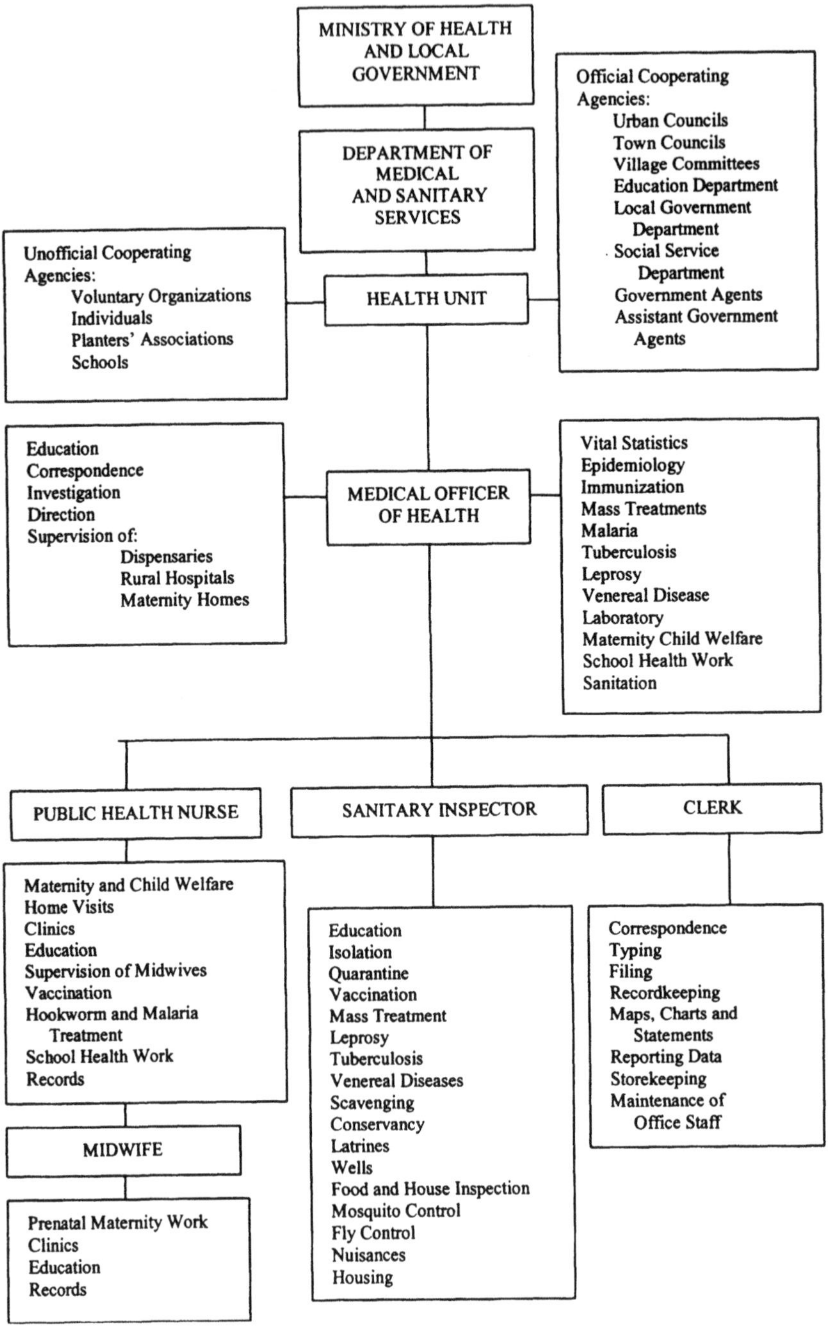

Figure 5.1: Organizational Structure of Health Units
Source: S.F. Chellappah and W.P. Jacocks, 1936 *A Guide to Health Unit* Procedure, Colombo, p. 50.

Notes

1. World Health Organization, *Primary Health Care. Report of the International Conference on Primary Health Care (*Geneva: WHO, 1978). Also, see WHO, *Targets for Health for All by the Year 2000,* ALMA-ATA (Geneva: WHO, 1978).

2. Daniel Taylor-Ide and Carl E. Taylor, *Just and Lasting Change: When Communities Own their Future* (Baltimore: Johns Hopkins University Press, 2002), 8-11.

3. Michael Durey, *The Return of the Plague: British Society and the Cholera 1831-2* (Dublin: Humanities Press, 1979), 155.

4. Quoted in Fraser Brockington, *World Health* (New York: Churchill Livingstone, 1975), 147.

5. *Ibid.*, pp. 147-8.

6. Victor Heiser, a pioneer health diplomat, began his career as a quarantine officer of the United States' immigration services, and later became the coordinating officer of the IHB for the Asia-Pacific region. Victor Heiser, *An American Doctor's Odyssey: Adventures in Forty-Five Countries* (New York: W.W. Norton, 1936), 37.

7. For a detailed study on the IHB involvement in Sri Lanka, see Soma Hewa, *Colonialism, Tropical Disease and Imperial Medicine: Rockefeller Philanthropy in Sri Lanka* (Lanham, MD: University Press of America, 1995).

8. Rockefeller Archive Center (hereafter RAC), 1926 W.P. Jacocks, *Preliminary Report on Health Units*, p. 4. RG. 462, Box 48.

9. RAC, 1926 *First Annual Report of the Health Unit Kalutara Badda*, p. 6. RG 5, Se. 3, Box 198.

10. *Ibid.*, pp. 8-13.

11. S. F. Chellappah and W. P. Jacocks, 1936 *A Guide to Health Unit Procedure*, Colombo, 7.

12. *Ibid.*, pp. 37-38.

13. RAC, 1932 *Report on Maternity Conference Held at the Office of the Director of Medical and Sanitary Services*, L. A. Prins, Deputy Director of Medical and Sanitary Service, p. 1. RG. 462.

14. RAC, 1932 *Note on Public Health Nursing in Ceylon*, p.1, RG. 462, Box 1.

15. RAC, 1939 *Ceylon Kalutara Health Unit Control of Soil Borne Disease*, p. 2, RG. 462, Box 1, Folder 12.

16. Chellappah and Jacocks, *A Guide to Health Unit Procedure*, 3.

17. *Ibid.*, 3.

18. RAC, 1932 *Maternity and Child Welfare, Annexure (M)*, p. 3. RG. 462.

19. Chellappah and Jacocks, *A Guide to Health Unit Procedure*, 3.

20. RAC, 1933 *Annual Report on Health Unit Work*, p. 3. RG. 5, Se. 3, Box 192.

21. RAC, 1934 *Ceylon, Kalutara Health Unit Designation for Two Years*, p. 1. RG. 1.1, Se. 462, Box 1.

22. RAC, 1935 *Ceylon, Kalutara Health Unit*, p. 1. RG.1.1, Se. 462, Box 1.

23. Government of Sri Lanka, *National Institute of Health Sciences, Kalutara, Sri Lanka* (no date).

24. International Bank of Reconstruction and Development, 1952 *The Economic Development of Ceylon*, Baltimore, Part II, p. 127.

25. Marga Institute, 1984 *Intersectoral Action for Health: Sri Lanka Study*, Colombo, p. 23; C. G. Uragoda, 1987 *A History of Medicine In Sri Lanka*, A Centenary Publication, Sri Lanka Medical Association, Colombo, p. 163.

26. M. D. de Silva, 1956 *Health Progress in Ceylon: A Survey*, Colombo.

27. RAC, 1932 *Notes on Public Health Nursing in Ceylon*, p. 1, RG. 1.1 Se. 462, Box. 1.

28. J. W. Bjorkman, "Health Policy and Politics in Sri Lanka. Development in the South Asian Welfare State," *Asian Survey*, (1983) 15:537-52.

29. World Health Organization, 1989 *Health Care in South-East Asia*, New Delhi, p. 191.

30. United Nations Development Program (UNDP) *Human Development Reports*, (New York: Oxford University Press, 2001).

31. Inger Karl and Michael Faust, "Global public goods and health: taking the agenda forward," *Bulletin of the World Health Organization*, (2001) 79:869-74.

32. *Ibid.*, 870. Also see, David Sanders and Mickey Chopra, "Globalization and the Challenge of Health for All: a View from Sub-Saharan Africa," in *Health Impacts of Globalization, Towards Global Governance*, ed. Kelley Lee (New York: Palgrave-MacMillan, 2003), 105-19.

33. T. Sandler, "International Public Goods: Strategies, Efficiency, and Institutions," in Inge Kaul, Isabelle Grunberg and Marc A. Stern (eds.), *Global Public Good: International Cooperation in the 21st Century*, (New York: Oxford University Press, 1999), 20-50.

34. Karl and Faust, "Global public goods and health," 869.

35. C. McCord and H. P. Freeman, "Excess Mortality in Harlem," *New England Journal of Medicine* (1990) 322:173-77.

36. Thomas McKeown, *The Role of Medicine: Dream, Mirage, or Nemesis?* (Princeton: Princeton University Press, 1979)

37. Arno Karlen, *Man and Microbes: Disease and Plagues in History and Modern Times* (New York: Putnam Book, 1995), 219.

38. Malcom G. Taylor, *Insuring National Health Care: The Canadian Experience.* Chapel Hill, NC: The University of North Carolina Press, 1990).

39. R. Jackson, *Issues in Preventive Health Care* (Ottawa, Science Council of Canada, 1986).

40. Soma Hewa, "Physicians, Medical Profession, and Medical Practice, in *Health, Illness, and Health Care in Canada,* (eds.) B. Singh Bolaria and Harley D. Dickinson, (Toronto: Thomson Learning, 2002) 55-81.

41. Jim S. Frideres, "Overcoming Hurdles: Health Care and Aboriginal People," in Bolaria and Dickinson ed., *Health, Illness, and Health Care in Canada*, 144-66.

42. Murray McQuigge "The Walkerton Disaster and Family Physicians," (editorials), *Canadian Family Physician*, (2003), 1-3.

43. Canadian Institute for Health Information, *Health Care in Canada* (Ottawa: Statistics Canada, 2005) 15-23.

44. Richard G. Wilkinson, *Unhealthy Societies, The Afflictions of Inequality,* (London: Routledge, 1996).

45. M. Jee and Z. Or, *Health Outcomes in OECD Countries: A Framework of Health Indicators for Outcome—Oriented Policy Making*, (Paris: Organization for Economic Cooperation and Development (OECD), 1999), 14.

Chapter Six

Of Mice, Vaccines and Men: The Yellow Fever Research Program of the Rockefeller Foundation in Colombia, 1932-1948

Paola Mejía

The understanding of yellow fever was radically transformed in the last years of the 1920s. The discovery of a susceptible animal in 1927 was rapidly followed by the development of new laboratory techniques, immunological tests, and 10 years later, by the availability of a safe vaccine for massive application. While the establishment of the yellow fever virus in an animal prompted research on the disease in laboratories in Africa, Brazil, the United States, England, Holland and France, the undisputed leader of yellow fever studies was the Rockefeller Foundation (RF). Its research program was oriented by its previous work in yellow fever control in Latin America, and the actual work in the field provided material for laboratory research. At the same time, field investigations could now be checked against laboratory results, and the combination of laboratory and field studies proved highly productive for the advancement of knowledge. This knowledge was readily translated into public health methods for disease control in tropical regions.

By the end of the 1930s, and in the context of these new technological developments, Colombia was at the forefront of yellow fever research in two state-of-the-art laboratories in Bogotá and in the eastern town of Villavicencio. Native and foreign scientists devoted their efforts to the study of the epidemiology and transmission dynamics of a recently discovered variety of yellow fever, the so-called *jungle yellow fever*. In this article I analyze why and how Colombian public health experts made the study of yellow fever a national priority and committed considerable resources to a cooperative research project with the Rockefeller Foundation, as well as some of the consequences this had in the development of public health and biomedical research in the country.

The Colombian yellow fever program was part of an international network of eight Rockefeller Foundation-supported laboratories encompassing Uganda, Nigeria and Brazil, with its epicenter in the Yellow Fever Laboratory in New York City. With half of the funding provided by the Colombian government and the other half by the Rockefeller Foundation, the program lasted ten years and concentrated exclusively in scientific research. A close look at the program shows that the RF supported the Colombian yellow fever program with an interest in gathering scientific information in order to develop strategies for the global control of yellow fever, but was careful not to involve itself in the control of yellow fever in Colombia. This approach differed from other RF sponsored yellow fever programs in Latin America, such as the one in Brazil. There, the RF was in charge of the National Yellow Fever Service and worked intensively in mosquito control. Fred L. Soper, head of the RF in Brazil and its representative for South America played a leading role in the promotion of laboratory work as a means to control yellow fever. To convince the nations of the Americas to embark in a complex and costly yellow fever research program, Soper constructed yellow fever as a public health menace of continental proportions requiring heavy investments and concerted international cooperation for its control, one that would use the laboratories in Brazil, Colombia and New York as reference centers.

For the Colombian public health authorities, the yellow fever program was a source of prestige for both the government and the nascent discipline of public health. The availability of a modern research laboratory that rendered services and furnished vaccines for free to neighboring countries was a matter of national pride for government officials in the era of high nationalism that followed a military conflict with Peru in 1932. The yellow fever program was also a symbol of the commitment of the liberal party, back in power in 1930 after half a century of conservative hegemony, with the health and welfare of the masses. Nevertheless, the RF involvement in yellow fever in Colombia curtailed cooperation in other health programs that were more relevant for the pressing public health needs of the country. The RF funding for the yellow fever program was out of proportion with any other health project in Colombia, and the government furnished 50% of its cost during all the period in which the program was active. The local manufacturing of yellow fever vaccine and mass vaccination of populations at risk were the most visible outcomes of the program, while the parallel research in entomology, virology and epidemiology deepened the scientific understanding of the disease and served as a training ground for a generation of Colombian biomedical researchers. Nevertheless, the tensions generated by basic research in a country with pressing public health needs led to a conflict between local health authorities and the RF that ended in the nationalization by decree of the yellow fever field laboratory in Villavicencio and the abrupt termination of the cooperative yellow fever program in 1948.

The RF role in the shaping of international public health and scientific research has been extensively studied in recent decades.[1] Latin America has not been absent from this discussion and there are excellent works about the role of its International Health Division (IHD) as an agent of modernization, or in other words, "Americanization" of Latin American public health and medicine.[2] The most recent studies call our attention to the interactions between the local and the international aspects of public health work. Solorzano's work for instance shows how the yellow fever program in Mexico helped the local government to secure control of a region where insurgence, not yellow fever, was the felt menace, and where the RF played an important role as an agent of American diplomacy.[3] Marcos Cueto, studying the Peruvian case, evaluates a yellow fever campaign that, though effective, was conducted in a vertical and authoritarian style that clashed with the local population.[4] Steven Palmer, analyzing the RF programs in Costa Rica, invites scholars to go past the implicit imperialism of public health programs in order to understand the ways in which local actors used the RF resources to strengthen pre-existing local initiatives and national public health.[5]

While the work of the RF in yellow fever in Brazil is well documented,[6] the Colombian case has been scarcely studied, excepting the work of Christopher Abel and studies that focus on the hookworm campaigns.[7] The yellow fever program of the RF was studied by Augusto Gast Galvis, a Colombian physician who trained in biomedical research through his work in this program and became a leading figure in yellow fever in the country. Galvis wrote a history of yellow fever in Colombia that surveyed all documented yellow fever epidemics in the country, the early jungle yellow fever studies, and devoted its last pages to acknowledge the importance of the RF in the research programs of the 1930s.[8]

The case of yellow fever studies in Colombia highlights some of the issues faced by contemporary international health as it confronts challenges such as the resurgence of old diseases, emerging infections, or drug resistance. While research is an important piece of the response to some health problems, the countries most heavily affected are also the poorest -as is the case of malaria in sub-Saharan Africa-, posing both ethical and technical questions about priorities in health and social investment. At the same time, past experience shows that the failure to incorporate research as a component of disease control efforts may hamper the effectiveness of public health interventions.[9]

Yellow Fever and the Rockefeller Foundation

While some European countries embarked in yellow fever research in laboratories both in Europe and in their tropical colonies in the twentieth century, there is no doubt that the leader in the field during the period of our consideration was the Rockefeller Foundation through its International Health Division (IHD). In fact, yellow fever would dominate IHD programs from 1916 on, and its reputation was gained mainly there, linking its name to the field of yellow fever research.[10]

While the original purpose of the IHD at the time of its creation had been to extend the hookworm work of the Rockefeller Sanitary Commission in the Southern states of the United States into Latin America, the Orient and the British Empire,[11] yellow fever was soon to be added to its program. This was due in part to the concern that the opening of the Panama Canal in 1914 could facilitate the transport of yellow fever to Asia in infected ships. Known to Africa, Europe and the Americas, yellow fever was never present in the East. Such an event could theoretically be the starting point of catastrophic epidemics, since the vector mosquito was present in Asia and its population was immense and susceptible. Yellow fever had been feared for centuries due to its spectacular symptoms (high fever, yellow skin, hemorrhages, blood vomit) and its high lethality: one in three victims would die from the disease. The appearance of yellow fever in port cities wreaked havoc in international commerce and terrified the population, giving rise to a complete interruption of normal life and business.

Confronted with this problem, Wickliffe Rose, director of the IHD, called into consultation the three foremost experts in yellow fever at the time: William Gorgas, Henry Rose Carter and Joseph White. General William Gorgas, whose sanitation of the Panama Canal with techniques partly developed in Havana at the beginning of the twentieth century allowed the construction of the Canal, was convinced that eradication of yellow fever was possible and convinced Rose of the worthiness of the enterprise. Carter thought that the risk was very small over that great distance—and Africa was much nearer in any case—but the time was opportune to sell the idea of eradication.[12] A meeting of an International Health Commission (IHC) was held in New York, with the presence of Drs. William Welch, Abraham Flexner, and John D. Rockefeller Jr. as Chairman and the case for yellow fever eradication in the Western hemisphere was presented by Gorgas, Carter and White. The Commission, after careful consideration, concluded that "the IHC is prepared to give aid in the eradication of this disease in those areas where the infection is endemic and where conditions would seem to invite cooperation for its control."[13] Thus the IHD of the RF committed to the eradication of yellow fever in the Americas. The yellow fever program was active for 34 years and 76 RF staff members were involved in it at one time or another. It cost almost $14 million, three times as much as was spent on any other single disease or objective.[14] From 1925 until the late 1930s, more than 50% of the budget for diseases was assigned to yellow fever.[15]

The Yellow Fever Commission was created in 1916 as part of the new commitment of the RF to yellow fever eradication. The expert consensus, condensed by Rose in the 1915 Annual Report, stated that

> the endemic foci [of yellow fever] are the seed-beds of infection; and if these seed-beds be destroyed, the disease will disappear from all other points. There are probably not more than five or six endemic foci all told, so the problem of eradication of yellow fever reduces itself to stamping it out at these five or six points.[16]

The Commission visited suspected endemic centers in Ecuador, Peru, Colombia and Venezuela. Later that year, it also visited the principal ports of Brazil, from Rio de Janeiro to Pará, concluding that the only endemic center in South America was in Guayaquil, Ecuador.[17] This theory came to be known as the Key Center Theory, the scientific foundation of eradication efforts.

In 1918, success in controlling yellow fever in Guayaquil, along with the announcement by Rockefeller scientist Hideyo Noguchi of the isolation of the causal agent, *Leptospira icteroides* (wrongly, as later proved), seemed a good omen for the eradication effort. By the end of 1924, the disease had disappeared from Mexico, Central America and Ecuador.[18]

By the mid-twenties, yellow fever was considered a vanishing disease. With manpower and resources freed from Latin America, the Foundation opened a new front in Western Africa. In 1925, the West Africa Yellow Fever Commission started to work with specific objectives in mind: to determine whether African yellow fever was identical with that of the Americas and whether the African conditions were favorable for a similar campaign of eradication.[19] In an exhaustive study, the commission failed to isolate Noguchi's *L. icteroides*. As laboratory work was handicapped by the lack of an experimental animal, the group concentrated in finding a susceptible animal. In 1927, they succeeded in infecting rhesus monkeys from India with yellow fever.[20] The development of an animal model opened an era of laboratory research in yellow fever and led to the isolation of the filterable virus and the final demise of Noguchi's *L. icteroides*.

By the time these discoveries were taking place, the IHD was shifting its emphasis toward scientific research. For Dr. Frederick Russell, who became Director of the IHD in 1928, the purpose of the Division's work was not "to make demonstrations which will lead ultimately to the enlistment of local agencies in the work", as Wickliffe Rose, his predecessor, had promoted, but to pursue scientific studies in public health.[21] Russell began to press for a central yellow fever laboratory in the United States where more thorough work could be carried out. In June 1928, two rooms of the Rockefeller Institute in New York became the Yellow Fever Laboratory.[22]

The Yellow Fever laboratory was to bring together the viruses of Africa and South America under one roof in New York, where they could be studied in detail by trained scientists, with state of the art equipment and supplies. New York was not only equidistant from the two areas of yellow fever prevalence, but was outside the zones of natural occurrence of the virus.[23] It was also the home of the Rockefeller Foundation.

New Tools for the Study of Yellow Fever

The Mouse Protection Test

With the isolation of the yellow fever virus, work at the Yellow Fever Laboratory of the RF in New York centered in the development of an efficacious vaccine.[24] Laboratory research received a boost when Max Theiler, at that time at the Harvard School of Tropical Medicine, found that mice were susceptible to intracerebral inoculation of yellow fever virus, and that human immune serum protected mice. This finding reduced enormously the cost of research and would grant him the Nobel Prize in medicine in 1951. Based on this findings,

Wilbur Sawyer and Wray Lloyd developed a "mouse protection test" that revealed the presence of specific antibodies to yellow fever in humans serum. The mouse was injected with blood from a person, and with the yellow fever virus afterwards. If the person had ever suffered from yellow fever, even without manifesting any symptoms, the antibodies would protect the mouse from infection. If not, the mouse would die from the infection. This test allowed the retrospective diagnosis of a yellow fever infection in humans, notwithstanding how long ago it had taken place. The test made possible immunological surveys that revealed past contacts of tested populations with the virus, allowing the mapping of endemic areas.[25]

The IHD began to use the mouse test in a widespread survey of the geographic distribution of immunity to yellow fever. Between 1931 and 1949, 95,000 human sera from all over the world were examined.[26] These surveys were not only of epidemiological interest; as Heather Bell has remarked, the existence of this protection test provided a vital tool precisely at the time when international health bodies were considering how to address the health threats arising from the newly dawning age of commercial air travel.[27] In October 1930, the Office International d'Hygiene Publique asked the RF to take charge of a survey of yellow fever endemic areas in Africa with the purpose of locating aerodromes outside the danger zone.

The survey started in West Africa, but was extended to the Belgian Congo by late 1932. As initial results were confusing, it continued eastward into Anglo-Egyptian Sudan. The specimens from Sudan tested in the New York laboratory gave startling results: 42% of them were positive for yellow fever immunity, in a country where a case had never been described. The IHD was eager to see a yellow fever laboratory opened in Sudan, but the government was wary of yellow fever research, and did not yield to pressure. Bell concludes that the real reason behind Sudan's government negative was that yellow fever, barely present in the country, was not a priority that the country could afford. Moreover, the building of a laboratory at that time was not feasible due to the local political circumstances of bacteriological research at the time.[28]

The Yellow Fever Vaccine

Research at the Yellow Fever Laboratory in New York led to the development of an effective vaccine in 1937, the 17D vaccine, using a modified strain of the yellow fever virus isolated from an African patient. The RF took on the manufacturing of large doses of vaccine. As a result of a plan set up to produce influenza vaccine should a pandemic arise, the Foundation had vaccine manufacture capabilities. Influenza didn't return, so the facilities stood idle. Commercial firms were not anxious to make yellow fever vaccine: there was no money in it, and they wanted $2 a dose. The RF could make it for two or three cents, including staff salaries. The Foundation decided to manufacture yellow fever vaccine as a contribution to the war effort, and with that purpose employed the flu lab.[29]

Viscerotomy

An additional tool for the study of yellow fever was the histopathological study of the liver of victims of yellow fever. Lesions of yellow fever in the liver were described as early as 1890, but it was up to a Brazilian experimental pathologist, Henrique da Rocha Lima, working in Germany, to publish in 1912 detailed descriptions using modern staining techniques. This finding was put to use in Brazil in the late 1920s, and a technology for obtaining liver specimens without an autopsy was perfected in 1930. It employed a very simple instrument that could be easily handled by people without medical training. The technique, known as viscerotomy, was applied to all patients that died from an acute febrile illness and allowed an exact postmortem diagnosis of yellow fever.[30]

In a few years, the laboratory had radically transformed the identity of yellow fever.[31] Instead of relying on clinical observation and experience, the scientific diagnosis of yellow fever now depended on the isolation of the virus and the study of tell tale lesions in the liver of yellow fever victims, tests that could only be done by experts in a handful of specialized laboratories. Past infection could now be detected through the mouse protection test.

Yellow Fever as a Continental Problem

As we saw above, from 1927 on the conceptualization of yellow fever had been changing at an accelerated pace, and the laboratory was now a central element in the study and control of the disease. Very soon, the epidemiology would need a revision as well. In 1928 there was an unexpected outbreak of yellow fever in Rio de Janeiro, the first in the twenty years elapsed since Oswaldo Cruz stamped out yellow fever from the city at the turn of the century. Between 1929 and 1932, four unexpected epidemics that did not fit the accepted key-center epidemiology occurred at widely separated points of South America in Venezuela, Brazil, Bolivia and Colombia.[32]

The origin of none of these epidemics could be traced, and neither in Brazil nor in Bolivia was the *Aedes aegypti* mosquito, considered the sole yellow fever vector, found. This type of yellow fever without *Aedes aegypti* in isolated areas with scarce population was described in 1935 as a separate entity, jungle yellow fever, by Fred L. Soper, head of the RF in Brazil.[33]

Fred L. Soper attended the Ninth Pan-American Sanitary Conference in Buenos Aires in November 1934 as an observer. In the conference, he gave a lecture entitled *El problema de la fiebre amarilla en América*, (The yellow fever problem in the Americas) where he presented the recent discoveries in Brazil, Colombia, Bolivia and Africa, and pressed for the acknowledgment of yellow fever as a continental problem, quoting his own intervention the previous year before the Academy of Medicine of Brazil:

> Brazil has recognized yellow fever as a national problem, but in the light of our present knowledge it must be considered an international problem. Discoveries in recent years in Brazil, Colombia, Bolivia and Africa demonstrate that for and adequate understanding of the problem, a complete survey, with the modern

> methods, is needed in all previously endemic regions. In South America it can be said that yellow fever must be considered a continental problem, linked to the interests of all its constituent nations . . . With the increase of transportation in many places, and an aerial service that allows to circle South America in 13 days (soon it will be possible to do the circuit in seven days, with a maximum of four days between ports), yellow fever in any point of South America concerns equally all nations of the continent.[34]

From his perspective, a "continental program" should be a collaborative effort based in: 1) unified norms for anti-larval measures and for the systematic collection of liver tissue from rapidly fatal febrile cases, and 2) the establishment of laboratory services for doing protection tests, liver examinations, entomological studies and research on possible animal hosts of the virus.

When Soper referred to yellow fever demonstration in his paper, clinical diagnosis would not suffice: in Valle de Canaán, Espiritu Santo, Brazil, yellow fever was identified by clinical observation, autopsy, mouse protection test and reproduction of the disease in monkeys inoculated with blood from one of the patients. In Bolivia, yellow fever was only the one proved by clinical examination, histopathological study of the liver and mouse protection test.[35]

As regional director for the RF in South America, Soper was in a unique position to move ahead his yellow fever scheme. In a conference before the School of Medicine at the National University in Bogotá in April of 1935, Soper announced the recommendations issued at the ninth Pan-American Sanitary Conference regarding yellow fever. One of them was the founding of special laboratories for the study of yellow fever. It had been agreed that "those countries which do not have specialized laboratories or institutes should make agreements with neighboring countries or with private institutions of international nature for the carrying out of the necessary investigations."[36] After a briefing on the epidemiological findings relevant to Colombia, and the public recognition of the early description of the epidemiological characteristics of jungle yellow fever by Colombian physician Roberto Franco, Soper announced that the Minister of Agriculture, the Director of the National Department of Health, and the Director of the National Hygiene Laboratory were considering plans for the organization of a yellow fever service at that laboratory. Soper presented the field of vaccine and sera production as a legitimate activity for the National Hygiene Laboratory, which would add to the efforts of the Yellow Fever Laboratory of the Rockefeller Foundation in New York and the Laboratory of the Brazilian Yellow Fever Service in Rio de Janeiro.[37]

The Rockefeller Foundation and Yellow Fever in Colombia

Before 1927: The traditional epidemiology of yellow fever

The RF appeared in the Colombian public health scene in 1916, when the government contracted the Foundation to survey foci of yellow fever. It was invited back in 1919 to survey hookworm infection. From then on, the Foundation had a continuous presence in public health activities in Colombia until the fifties, diminishing only in 1951, when the Foundation shifted its priorities to agriculture and nutrition.[38]

The RF commission that visited Colombia in 1916 was the newly created Yellow Fever Commission that had studied the "key centers" in South America. It surveyed the towns of Puerto Berrío, Muzo and Cúcuta and stated that there was no endemic yellow fever in the country. Part of the material they studied as part of their investigation consisted in the statements of Drs. Henao, Toro Villa and Montoya, three of the most prestigious physicians of the Medellín scientific community, identifying the disease in Puerto Berrío as yellow fever. One of the fatal cases had been autopsied by Dr. Henao, and portions of the liver obtained in the necropsy were sent to Dr. Balfour, a world-renowned authority in tropical diseases, and Dr. Stevenson of the Wellcome Laboratories of Research, in London. The negative laboratory tests, and the opinion of the American commission, led to the abandonment of the diagnosis of yellow fever.[39]

When in 1923 a yellow fever epidemic was reported in Bucaramanga (Santander), the Colombian government requested the advice of the RF, and it sent Dr. J.H. White, Assistant Surgeon General of the United States Public Health Service, Dr. Oliver Pothier, a bacteriologist from New Orleans, and Dr. Wenceslaus Pareja, Director of Health of Ecuador as the IHD representatives. The commission reported the presence of yellow fever, and following this report, the RF was asked to cooperate in the control and elimination of the disease.[40] With that purpose, Dr. Henry Hanson and Mr. L.H. Dunn were designated by the RF to conduct a campaign against the mosquito.[41] Dr. Henry Hanson came fresh from his yellow fever campaigns in Guayaquil in 1918 and in Perú from 1919 to 1921.[42]

One of the first tasks of this new commission was to investigate the rumors of suspicious fevers in other areas of the country. Confronted with the Rockefeller expertise, which had denied that the epidemic taking place in Ebéjico (Antioquia), and christened Ebéjico's fever, was yellow fever, a physician wrote in 1924 in the *Revista de Higiene*, voicing the frustration of some of his Colombian colleagues:

> We know that it is a scientific assault to contradict the opinion of the American Mission; nevertheless, the brief permanence of the technicians, who only stayed for 24 hours, the fact that they didn't get to Guayabal, which has always been the diffusion focus of the endemo-epidemic, the insufficient capture of mosquitoes for making it a serious survey, greatly undermine the statements of the Americans as authoritative criteria.[43]

For many years, the RF staff had dismissed yellow fever cases that did not adjust to the accepted theory, overlooking the opinion and expertise of local physicians, as has been described in the case of Brazil.[44]

Yellow Fever in Muzo: a Long Story

In the emerald mining region of El Muzo, about 150 Km north from Bogotá, yellow fever outbreaks had been reported by local doctors at frequent intervals since 1885. In 1906-07, an outbreak of a suspicious disease occurred in Muzo and was investigated by Drs. Roberto Franco, Gabriel Toro Villa and Jorge Martínez Santamaría, leading experts in tropical disease in the country, who performed autopsies, made blood examinations and collected mosquitoes. Dr. Franco concluded that the outbreak of Muzo in 1907 was yellow fever. Franco highlighted the fact that yellow fever in the area had certain particularities: it was contracted in the forest and not in the neighborhood of the houses; it was probably transmitted by several mosquitoes, and inoculation took place during daylight hours, which were spent by the workers in places where the transmitting mosquitoes predominated.[45]

Franco had clearly described the epidemiology of jungle yellow fever in his report, but his results were overlooked at the time, since they did not adjust to the accepted knowledge of *Aedes aegypti* as the sole vector. When yellow fever made its appearance again in the Muzo region in 1930, the epidemic was investigated the following year by Dr. Moreno Pérez, an ex RF-fellow, and other colleagues, with the aid of the new techniques developed by RF scientists.[46]

Moreno Pérez took samples of blood and sent them to the Rockefeller Institute in New York, where they were to be subjected to the recently developed mouse protection test and yielded positive for yellow fever. Being this test in its early days, it was decided to repeat the testing the following year, and in 1932, Drs J. A. Kerr from the RF and Dr. Luis Patiño Camargo, from the Departamento de Higiene, visited the region. Although no *Aedes aegypti* were found, the protection tests in 1932 confirmed the work of the previous year. Kerr and Patiño concluded that yellow fever was endemic in the rural areas of Muzo, and set to publish the results. Fred Soper pronounced himself against the conclusion, and the publication in English was cancelled, although the objections arrived too late to prevent the publication in Spanish in the local hygiene journal.[47]

When in 1934 a series of cases appeared among workers in the Muzo mines, Dr. Soper personally visited the mines, accompanied by Drs. Bevier and Patiño. Later that year, suspicious cases of a pernicious fever were noted by Dr. Jorge Boshell, a Colombian physician working in Restrepo, in the Eastern Planes. The cases appeared in an isolated rural area, among field workers from the piedmont. An autopsy and the demonstration of previous infection through protection tests determined that the disease was yellow fever. This outbreak, in an area where *Aedes aegypti* was not found, and without any possible contact with known yellow fever foci, was highly interesting for Colombian and RF scientist alike.

In a speech before the School of Medicine in Bogotá in 1935, where he announced the plans for the establishment of a yellow fever service at the National Hygiene Laboratory, Soper publicly acknowledged that Dr. Franco had described jungle yellow fever thirty years before. He also recognized that he had opposed the publication in English of the research results of Drs. Kerr and Patiño study in Muzo in 1934.

In any case, Soper was highly effective in promoting his extensive yellow fever program, and the following year, the Departamento de Higiene and the RF signed a contract by means of which the Section of Special Studies of the Departamento de Higiene came into being. The contract, considering that yellow fever existed in various regions of Colombia with unexplained characters that should be fully researched, and that malaria was an important health problem in Colombia, created the Section Eight of the National Health Department, devoted to investigations in yellow fever and malaria.[48] Resources were not equally generous for both diseases. Each party contributed US$30,000 for yellow fever studies, and US$5,400 for malaria studies. Funds for yellow fever were 5.5 times higher than those for malaria, although only a handful of cases of yellow fever had been reported: three in 1934, seventeen in 1935 and sixty-six in 1936, the year with the highest number of reported cases in the country.[49] In contrast, the incidence of malaria in the country was estimated at 3.6 million cases and 18,000 deaths annually in 1942, in a population of less than ten million.[50]

The Yellow Fever Program in Colombia: Responding to Hemispheric Needs

Before the establishment of the cooperative program in yellow fever, the RF work in Colombia had developed around hookworm control. By the time the RF stepped into yellow fever work in Colombia, the backbone of its activity in the country had been hookworm work. Since its initiation in 1919, the hookworm work had evolved into health education and rural sanitation. After many setbacks, the program was apparently starting to give fruits. George Bevier, the RF representative, was fully committed to rural health, an approach that was no longer favored by the IHD.[51] The Foundation was not pleased with Bevier's orientation, that stressed soil sanitation, the control of rural water supplies, farm sewage disposal, garbage disposal and the control of flies, the corralling of domestic animals, and, to a limited extent, the destruction of domestic mosquito breeding places, and questioned his lack of attention to treatment.[52] Bevier had also advocated for a malaria program for Colombia for several years, and by 1934 a modest project was being started.

The period of the signing of the new cooperative agreement for yellow fever work was crucial for political developments in the field of public health. Since 1930, the Liberal party, back in power after three decades of conservative hegemony, was progressively attempting to incorporate popular demands into the governmental project. The state was expanding its capacity of action in hygiene matters, trying to include traditionally excluded social groups like rural peasants and workers. Rural peasants entered the political constituency as the government proposed to integrate the rural population into the national economy both as producers and as consumers. The practical way to do so was to support the cultivation of public lands by *colonos*, whom in this way became the protagonists of national development.[53]

By the mid-thirties, there was a debate in the National Hygiene Department around different alternatives for the provision of integrated sanitary and health services in the rural areas. The government acknowledged the inadequacy of single-oriented campaigns in regions that had complex and multiple health problems, and created the rural sanitary units as a mean to provide access to public health and services. The "defense of the public health and the stimulation of the biological capacities of the country" were organized around Sanitary Units, Rural Commissions and Sanitary Offices, depending on the economic capacities of the regions.[54] The National Hygiene Department began the establishment of these units in 1936, an effort supported by the RF in some areas of the country.[55] Thus, many of the efforts in public health organization tended to decentralization and integration of public health activities.

Moreover, the National Department of Hygiene was reorganized early in 1936. The old hookworm section, became the Rural Sanitation Section and included direction of the 45 new health units, "if and when they start to function", commented RF's Dr. Paul, and the new cooperative services in yellow fever were organized as the Section 8, Special Studies Section.[56]

The RF had begun to explore the possibilities of a yellow fever research program in Colombia several years before the agreement. By August 1933, the RF was instructing his representative in Colombia about the desire of the IHD director to have an investigational service for yellow fever in the country, which would serve as a centralized entity for research in Perú, Bolivia and the Amazon Basin, and possibly also Venezuela and Guiana, promoting economy and efficiency for the RF goals.[57] A month later, Soper visited Colombia with the purpose of "calling attention to the need of further intensive studies of the distribution of yellow fever" and to outline a program for such studies similar to the one already going forward in Brazil and Bolivia.[58] The Director of Hygiene at that time happened to be an ex-fellow of the RF, and was willing to cooperate.[59]

A year later, when Soper visited Muzo and Caparrapí, where yellow fever had been reported recently, Dr. Rickard from the RF arrived to organize a viscerotomy campaign and begin an investigational program. He was shortly followed by Dr. J. Harland Paul and Dr. Paulo Azevedo Antunes, a medical entomologist from Brazil.[60]

The original plan called for the creation of a specialized yellow fever section in the already existing National Hygiene Laboratory. By 1935, Dr. Bernardo Samper had recently returned to the direction of this institution. He had been one of the two original founders of the laboratory, opened in 1917 and purchased a decade later by the government, who converted it into a National Hygiene Laboratory by RF recommendation. Samper was excited with the idea and quite anxious to organize and develop facilities for the laboratory study of yellow fever in the country.[61]

Soper wrote to the Director of Hygiene, outlining a program for yellow fever research that included the organization of a viscerotomy service, immunity surveys, investigation of suspect epidemics, organization of epidemiological studies, including determination of important vectors and possible vertebrate hosts, and the organization of field laboratories. He made clear, nevertheless, that "it is not to be expected that the Yellow Fever Service take any part in vector control in urban areas, which are under the Hygiene Department." A yellow fever section in the National Hygiene Laboratory was to be created, but although the RF would give some initial support, the budget for laboratory animals should be a responsibility of the National Laboratory, not the Foundation.[62]

In August 1936, Austin Kerr was sent to Colombia to organize the Yellow Fever Section. He replaced Rickard, who did not get on with Bevier.[63] Presumably, the tensions between them were the result of the clash of two very different outlooks on public health, Bevier representing the old goals of laboriously building health services capacity, while Rickard embodied the modern orientation of the IHD toward laboratory research. Kerr arrived with instructions to set up a research institute and a field laboratory to carry out transmission experiments with various arthropods and to make susceptibility tests on various animals. Soper called this an "Epidemiological Institute."[64]

Meanwhile, the establishment of viscerotomy posts had been advancing since 1934. The securing of liver specimens was backed by the passing of a law that forbade the burial of corpses without a certificate from the viscerotomist. At its peak, viscerotomy posts would be close to 114 around the entire country.

Kerr started to work on the organization of the Section of Special Studies. The National Hygiene Institute handed over a space in its premises for the provisional establishment of the yellow fever equipment. While early plans contemplated a yellow fever laboratory as part of the National Hygiene Laboratory, Kerr envisioned the yellow fever section as a national medical research institute. At some point, he contemplated its location in the new headquarters of the National University, to Samper's dismay, who lamented the growing tendency towards independence of the yellow fever laboratory.

Discussing his ideas with government officials, Kerr was questioned about the way in which his project related to the Institute of Tropical Diseases that the National University School of Medicine wanted to set. For Kerr, there was no conflict whatsoever, as he always saw the National University's project as an attempt to provide instruction facilities.[65] Many times -and by several prestigious politicians and physicians- were the RF representatives asked for support for this project, but it was never contemplated seriously. The Colombian government also asked the RF repeatedly to step into studies of bartonellosis, which had been uncovered as a public health problem in the south of the country, but Sawyer's instructions were clear: limit work to yellow fever.[66]

Who Should be in Charge of Vaccination?

Once the decision was made to construct a new building in Bogotá for the yellow fever laboratory, vaccination started to take increasing attention. Quietly, the Section of Special Studies had been providing vaccination by special requests, and in areas where there were flare-ups of yellow fever. On examination, most of the early vaccinated were foreign personnel of oil companies. The Section also provided vaccines for road and railroad construction workers. Vaccine was supplied by the New York or Rio laboratories. In Brazil, a vaccine trial was under way for the design of techniques of mass vaccination.

Kerr considered it entirely logical that the National Health Institute should proceed with the commercial preparation of vaccine, working in close cooperation with the Special Studies Section. Vaccination was to be done ideally by the personnel of the existing Sanitary Commissions of the National Department of Hygiene.[67] A year later, he was favoring a specialized campaign as the best way to organize yellow fever vaccination in the country. But a yellow fever campaign did not sell well with Colombian health authorities. Commenting on their attitude, Kerr wrote in his diary:

> . . . it seems to me to be as great an error to attempt to include all phases of public health work into health units as it is to conduct all public health work by entirely specialized campaigns. In Colombia at present the enthusiasm is all for health units and any specialized campaign has hard going. This is of importance in regard to yellow fever not only because of the antistegomya campaigns but also on account of the yellow fever vaccination campaign which for the present at least, I believe can best be organized on the basis of a specialized campaign.[68]

In any case, things did not proceed fast enough for the interests of the RF. To expedite the process, Dr. John Bugher, at that time assistant professor of the Department of Pathology at the University of Michigan, was appointed to the staff of the International Health Division and sent to Colombia. Dr. Theiler and Dr. Bugher were to work in Colombia for six months, along with Kerr, and return to New York. They job was to get vaccine manufacture in Bogota and research in Villavicencio going.[69]

A year later, the course of the vaccination program was still undecided. The Foundation was eager to devote exclusively to research, but vaccination manufacture and application stood in its way. Soper thought, like Kerr, that vaccination should be carried outside of the Special Studies Organization, and vaccine produced by the National Hygiene Institute.[70] But Sawyer, director of the IHD, thought otherwise:

> As you know, I have been much against our getting started on the control of yellow fever at the expense of the International Health Division and had hoped that control measures could be financed entirely by the Government while we devoted ourselves to investigation. I should, however, like to make an exception both here [Colombia] and in West Africa of the introduction of vaccination . . . I would not, however, wish to have the Division pay the major part of the control work through vaccination, if it can be avoided. As for the epidemiological study, I have the impression that it may be necessary for us to carry a large portion of the burden and possibly all of certain phases. I have no objection whatever to the Government's taking as much part as we can induce them to take. I would not, however, wish to see progress made at a slow pace for lack of reasonable amount of funds.[71]

Dr. Hugh H. Smith, who had worked at the Rockefeller Institute in New York with W. A. Sawyer and Max Theiler in the development of the 17D vaccine, and also in the Rio lab in vaccine production, was transferred to Colombia to get the manufacture started. The IHD had finally yielded to the passive resistance of the Colombian government to embark in the vaccination program.

The Special Studies Section always felt that vaccine application—not to mention epidemics investigation—would take away precious time for research into the basic mechanisms of jungle yellow fever, the real interest of the RF. But the Section couldn't say no: Colombian government was contributing 50% of the funds for the yellow fever program, which included, among others, vaccination and epidemics investigation. So mass vaccinations started at the end of 1938, organized by the Section with doses supplied by New York and Rio. At this time, Brazil vaccinations had reached a million people.

The Villavicencio Laboratory

The Villavicencio field laboratory was ready by mid-1938, and the Bogota yellow fever laboratory, about a year later. Bogota manufactured its first vaccine doses in 1940. By 1946, the Section—whose name was changed to Carlos Finlay Institute in 1945—had vaccinated a million people.[72] While much less visible, the bulk of the work in yellow fever research was carried out in the Villavicencio laboratory. The Colombian physician Jorge Boshell, who had called the attention of public health authorities to the presence of yellow fever in the Eastern plains piedmont in 1934, joined the staff of the Special Studies Section and worked continuously at the Villavicencio lab until 1946, directing research there in cooperation with the RF scientists that worked there temporarily, as well as with the Colombian physicians working in yellow fever. Important figures in Colombian biomedical research such as Augusto Gast Galvis, Ernesto Osorno, Manuel Roca García and Santiago Rengifo learn their trade in the yellow fever program. Boshell was perhaps the most notable character in this period; equally at home in the field and at the laboratory, he traversed the most secluded regions of the country in search of the yellow fever virus. In one of his trips to Santander, in the northeast of the country, he walked more than 200 kilometers, with fruit as his only sustainance at times.[73] He made one of the most useful discoveries for the study of jungle yellow fever: the *Hemagogus* vector, hardly found at ground level—a fact that greatly limited the early studies—was abundant in the forest canopy. Capture stations were thus devised at different tree altitudes, using the concept of vertical distribution of vectors.

At the Villavicencio laboratory, mosquitoes and other insects were captured and their habits studied both in the lab and at field observation stations. Different mammals and other animals were studied for yellow fever susceptibility, and several were found. The virus was isolated both from primates and mosquitoes, and transmission was obtained in the laboratory. The *Haemagogus* mosquito proved to be the most important vector for jungle yellow fever. Immunity distribution of yellow fever was studied through the mouse protection test in the most remote areas of the country, were also viscerotomy posts were established.

Withdrawal of the Rockefeller Foundation

By 1941, with yellow fever cases growing ever rarer and with an ample understanding of the epidemiology of the disease, the RF was pressed by the Colombian government to shift its action to other public health problems. The advent of the Second World War also dictated new research priorities at the RF laboratories. Nevertheless, the yellow fever program was maintained until 1945. After that year, the Villavicencio laboratory was reformed to accommodate the study of the natural behavior of the *anopheles* mosquito, the malaria vector.[74] To date, 26 different species of *anopheles* mosquito had been identified in the area, a region with one of the most rich mosquito fauna in the world.[75] In charge of this program was RF biologist Marston Bates. After working in Albania and Egypt, Bates was forced to leave the Mediterranean due to the outbreak of war, and his interest in tropical ecology led him to choose the Villavicencio station for his work.[76] In 1940 on Bates started to work in yellow fever, but in 1945, by indication of the RF, he started the *anopheles* program.

By 1946, as the RF was winding up its yellow fever work, plans were made to hand out the Bogota and Villavicencio laboratory to the government, but the volatile political situation of the country and the chaos in public health administration would mean that the laboratory so carefully built through a decade was bound for disaster. The RF decided to support it until it could be handed to someone competent. It was, in effect, supported for two more years. Now that the RF was planning to withdraw, its representative wanted Dr. Samper, director of the National Hygiene Laboratory, to incorporate the Yellow Fever laboratory into its own laboratory. The situation of the initial years in yellow fever activities reversed: now, Samper was extremely reluctant to receive it. Having trouble with government finances, he would have even more with this added responsibility.[77] Meanwhile, the Villavicencio laboratory continued its work in *anopheles* mosquito ecology. With World War II, malaria gained prominence.

The Colombian health authorities were not pleased with the plans for the Villavicencio laboratory. The town and its surrounding area had grown extensively in a few years, malaria was a visible problem and the people were demanding governmental intervention. The tension between research and control is illustrated by this discussion between Rolla B. Hill, the RF representative, and Hernando Rey Matiz, an ex-RF fellow, now director of the Malaria Service:

> Dr. Rey came in . . . principally to start arguing about our proposed malaria program. He wants, and has partly convinced Dr. Bejarano [head of the recently created Ministry of Health] while I have been away, that the RF program should be completely subordinated to his section, as far as malaria is concerned, and undertake investigation of malaria problems which are of immediate interest to him and his campaign for malaria control. His problem is urgent, and he wants immediate results; he is not interested in the long range study of anophelines, etc., such as we propose. He says the government money can be better spent in the things he is interested in. I told him plainly that if that were the case we would either withdraw entirely from contemplation of a malaria

> program, or support our program entirely with RF funds; that we were not interested in what he proposes, partly because we think it is time to get some fundamental data, partly because we do not have personnel.[78]

In the end, the government agreed to support the lab. But Hill was not pleased, for he would have preferred to support the lab exclusively with RF funds.[79] For all its work and the talented scientists that worked at one time or another in Villavicencio, in 1948, when Marston Bates tried to secure support for the continuation of basic research in the laboratory from the scientific elite of Bogotá, he found out that very little was known about its existence and the work it carried out among the medical and scientific circles in Bogotá.[80]

In early 1948, the RF representative found out through the newspapers that a new law had been passed, creating the *Roberto Franco Institute of Tropical Diseases*, whose headquarters would be located at the Villavicencio laboratory.[81] Upon further inquiry, he found out that beyond the personal conflicts between some Colombian physicians and Bates, in charge of the Villavicencio lab, the Minister was convinced of the righteousness and necessity of his cause. Public opinion in the district involved had forced him in his capacity as minister to utilize the existing facilities in the area for some type of control work as well as investigation and he wished to arrange plans so that both works could proceed simultaneously. The new activities would consist of some practical field studies directed toward a further elucidation of malaria vectors of the different localities as well as control operations, effected largely by painting the interior of houses with DDT, recently discovered. Thus, the incompatible goals of maintaining an area suitable for pure research with biology uninfluenced by control measures and, on the other, necessary control work directed toward "saving lives", were to coexist in the same small building.[82] Unsurprisingly, the deal did not work, and ultimately, Bates left the country and the RF withdrew from research in malaria.

Conclusion

The RF years of work in yellow fever in Colombia were a major force in the shift of Colombian medicine from a European, mainly French and clinical orientation, to the modern methods of American public health and medicine. The influence that the RF work exerted in a young generation of Colombian doctors is usually portrayed as a beneficial influence that supported the formation of a team of Colombian researchers in the field of yellow fever and other tropical diseases, and the period as one of the most actives in medical research in Colombia.[83]

The formation of a somewhat reduced group of medical researchers was a happy outcome of the RF yellow fever program. These researchers were highly specialized and formed the core of subsequent entomology and virology studies in the country, mainly in arbovirus. Physicians like Jorge Boshell, Ernesto Osorno, Augusto Gast-Galvis and Hernando Groot trained several generations of students in tropical medicine. Nevertheless, these researchers were trained because they were needed for the work of the RF, not because there was an explicit attempt to form a select group of researchers.

When the public health impact of the yellow fever work is presented, it is generally done in terms of number of vaccinations, or decline of number of cases. The number of vaccinations is an irrelevant parameter; the people easier to vaccinate are those with no chance whatsoever of encountering the yellow fever virus. Vaccination, even in remote regions and isolated towns, tended to reach those with less risk, for the truly exposed were interned in the forest as woodcutters, or perhaps as miners or road constructors, severed from contact with denser populations. That is the reason why yellow fever cases would continue to appear in areas were vaccination campaigns had been persistent.

In terms of cost, Colombia paid a very high price, one that perhaps was unaffordable when taking into consideration the pressing health needs of the country. After Brazil, who received 53% of all RF yellow fever expenditures (excluding scientists salaries), which were close to $10 million, Colombia was the second largest recipient of RF aid in yellow fever, but a second distant, with a 6.4% of total expenditures ($616,989.30)[84]. Roughly, Colombia's share of the yellow fever cooperative service was 50% during the entire duration of activities, so the country had paid more than $300,000 that went mainly to support the research of the RF.

It is clear that there were different options for the setup of the yellow fever cooperative activities; the laboratory in Bogotá could have been part of the National Hygiene Laboratory, in which case this institution could have been strengthened in its research capabilities, or it could have helped to the development of a research laboratory in the School of Medicine of the National University. The long held dream of many Colombian physicians and politicians of creating the Institute of Research in Tropical Diseases never received RF backing, even though the yellow fever laboratory could have been seen as a starting point.

Without doubt, the complex politics of Colombian public health in the period played an important role in the decision of the RF to "go alone" in its yellow fever project. But perhaps more important was the anxiety of the RF to secure scientific facts about yellow fever, facts that would serve as basis for the development of new control methods that guaranteed that north America would remain free of the disease, more so in the face of World War II. Thus, the RF paid scant attention to the long-term sustainability of the cooperative yellow fever service. This is noteworthy when contrasted with the Foundation policy of working in such a way that countries would be capable and willing to take over the projects initiated in cooperation after a reasonable time.

Three years after the incident at the Villavicencio laboratory that brought an abrupt end to Marston Bates' malaria research project, Hugh Smith, reflecting on the Foundation work in Colombia, told Lewis Hackett—who was at the time working in a history of the IHD—that when yellow fever was found in Colombia,

> all other work was abandoned as elsewhere . . . The feeling was that we were interested in our problems and not theirs. One reason they got rid of Bates was that. With the exception of yellow fever, Colombia has been a failure, with difficult communications and much sectionalism. Probably we are mainly to blame, like in Brazil where FLS [Fred L. Soper] took no interest in public health.[85]

Acknowledgements

Research for this paper was supported by a generous grant from the Rockefeller Archive Center through the Scholar-in-Residence Program and by its wonderful staff. Special thanks go to Thomas Rosenbaum, Mindy Gordon and Robert J. Battaly for their expert assistance in locating relevant sources, an to its director Darwin H. Stapleton. This paper benefited enormously by the comments of the editorial committee of *Dynamis*, where a version of this work in Spanish appeared in Vol. 24, 2004.

Notes

1. The public health work of the Rockefeller Foundation has been explored by historians from different perspectives; while early studies criticized the imperialist overtones of many RF interventions, for instance Saul Franco, "The Rockefeller Foundation's antimalarial program in Latin America: donating or dominating?" *International Journal of Health Services* 13.1 (1983): 51-67; Richard Brown, *Rockefeller Medicine Men: Medicine and Capitalism in America* (Berkeley: University of California Press, 1979), more recent studies focus on the analysis of local responses to American public health models. Among recent studies are John Farley, *To Cast Out Disease: a History of the International Health Division of the Rockefeller Foundation (1913-1951)*, (Oxford: Oxford University Press, 2004); Shirish N. Kavadi, "'Wolves Come to Take Care of the Lamb': The Rockefeller Foundation's Hookworm Campaign in the Madras Presidency, 1920-1928." In Esteban Rodríguez Ocaña ed., *The Politics of the Healthy Life: An International Perspective*, (Sheffield: European Association for the History of Medicine and Health Publications, 2002), 89-111; Ilana Löwi; Patrick Zilberman, "The Rockefeller Foundation and the Biomedical Sciences," *Studies in History and Philosophy of Biological and Biomedical Sciences* 31.3 (2000) (special issue); Paul Weindling, "Philanthropy and World Health: The Rockefeller Foundation and the League of Nations Health Organisation," *Minerva*, 35.3 (1997): 269-281; GIANNULI, Dimitra. "Repeated Dissappointment": The Rockefeller Foundation and the Reform of the Greek Public Health System, 1929-1940. *Bulletin of the History of Medicine* 72.1 (1998): 47-72; Soma Hewa, *Colonialism, Tropical Disease and Imperial Medicine:Rockefeller Philanthropy in Sri Lanka* (Lanham, MD: University Press of America, 1995).

2. Marcos Cueto, ed. *Missionaries of Science: The Rockefeller Foundation in Latin America (*Bloomington: Indiana University Press, 1994); Marcos Cueto, "The cycles of eradication: the Rockefeller Foundation and Latin American public health, 1918-1940" in Paul Weindling ed. *International Health Organizations and Movements, 1918-1939* (Cambridge: Cambridge University Press, 1995), 222-243; Ann-Emanuelle Birn and Armando Solórzano, "Public Health Policy Paradoxes: Science and Politics in the Rockefeller Foundation's Hookworm Campaign in Mexico in the 1920s," *Social Science and Medicine* 49 (1999): 1197-1213; Steven Palmer, "Central American Encounters with Rockefeller public health, 1914-1921," in *Close Encounters of Empire: Writing the Cultural History of U.S-Latin American Relations* (Dirham: Duke University Press, 1998), 311-332.

3. Armando Solorzano, *¿Fiebre dorada o fiebre amarilla? La Fundación Rockefeller en México, 1911-1924* (Guadalajara : Universidad de Guadalajara, 1997).

4. Marcos Cueto. "Sanitation from above: yellow fever and foreign intervention in Peru, 1919-1922," *Hispanic American Historical Review* 72.1 (1992): 1-22.

5. Palmer, "Central American Encounters with Rockefeller public health, 1914-1921."

6. Jaime L. Benchimol (coord.). *Febre amarela: a doença e a vacina, uma história inacabada* (Rio de Janeiro: Editora Fiocruz, 2001); Ilana Löwi, "What/who should be controlled? Opposition to yellow fever campaigns in Brazil, 1900-1939," in Andrew Cunningham and Bridie Andrews eds., *Western Medicine as Contested Knowledge* (New York: Manchester University Press, 1997), 124-146; Ilana Löwi, "Epidemiology, Immunology and Yellow Fever: the Rockefeller Foundation in Brazil, 1923-1939," *Journal of the History of Biology*, 30.3 (1997): 397-417.

7. Christopher Abel, "External Philanthropy and Domestic Change in Colombian Health Care: The Role of the Rockefeller Foundation, ca. 1920-1950" *Hispanic American Historical Review* 75.3 (1995): 339-376; Mónica García and Emilio Quevedo, "Uncinariasis y café: los antecedentes de la Fundación Rockefeller en Colombia: 1900-1919" *Biomédica* 18.1 (1998): 5-21.

8. Augusto Gast Galvis, *Historia de la fiebre amarilla en Colombia* (Bogotá : Instituto Nacional de Salud, 1982).

9. J. A Nájera, "Epidemiology in the Strategies for Malaria Control" *Parassitologia* 42.1-2 (2000): 9-24.

10. G.K. Strode, "Memo for Scientific Directors Meeting," October 26, 1944. RFA, RG 3.1, Series 908, Box 4, Folder LWH Notes for IHD History, v.III., 1215.

11. John Farley, "The Rockefeller Foundation: the Rusell Years," in Paul Weindling ed. *International Health Organisations and Movements, 1918-1939* (Cambridge: Cambridge University Press, 1995), 203-221 (204).

12. Lewis Hackett, Manuscript, Chapter 5, Running Start, p. 58. RAC, RG 3, Series 908, Box 5, Folder 32.

13. *Ibid.*, 61.

14. *Ibid.*, 56.

15. Farley, *To Cast Out Disease*, 88.

16. Rockefeller Foundation, Annual Report, 1915, 71.

17. Andrew J. Warren, "Landmarks in the Conquest of Yellow Fever," in George K. Strode ed., *Yellow Fever* (New Cork: McGraw-Hill Book Company, 1951), 14.

18. *Ibid.*, 17.

19. L. W. Hackett, "Epidemiology of Yellow Fever," RAC, RG 3, Series 908, Box 4, Folder 27.

20. Adrian Stokes, J.H. Bauer, and N. P. Hudson, "Transmission of Yellow Fever to Macaccus Rhesus: A Preliminary Note" *Journal of the American Medical Association* 90 (1928):253-254.

21. Farley, *To Cast Out Disease,* 203.

22. *Ibid.*, 213.

23. Lewis Hackett, "History-Lewis Hackett notes-Yellow Fever," RAC, RG 3, Series 308, Box 4, Folder 27.

24. Lewis Hackett, LWH Notes for IHD History, Vol. III, Pages 991-1544. RAC, RG 3, Administration, Program and Policy, RG 3.1, Series, 908, Box 4.

25. Löwi, "Epidemiology, Immunology, and Yellow Fever."

26. Hugh H. Smith, "Controling Yellow Fever," in Strode ed., *Yellow Fever*, 570.

27. Heather Bell, *Frontiers of Medicine in the Anglo-Egyptian Sudan, 1899-1940*, (Oxford: Clarendon Press, 1999), 168.

28. *Ibid.*, 163-167.

29. Lewis Hackett, "Notes-Yellow Fever," RAC, RG 3, Series 908, Box 4, Folder 27.

30. Fred Lowe Soper, *Ventures in World Health* (Washington: Pan American Health Organization, 1977), 151.

31. For a detailed study of laboratory and identity of infectious disease, see Andrew Cunningham, "Transforming Plague," in Andrew Cunningham and WILLIAMS, Perry Williams eds., *The Laboratory Revolution in Medicine*, (Cambridge: Cambridge University Press, 1992), 209-244.

32. Soper, *Ventures in World Health*, 169.

33. *Ibid.*, 168.

34. Fred L. Soper, "El problema de la fiebre amarilla en América," *Boletín de la Oficina Sanitaria Panamericana*, marzo (1935): 212.

35. *Ibid.*, 206.

36. Fred L. Soper, *Rural and Jungle Yellow Fever: A New Public Health Problem in Colombia* [Lecture given before the Faculty of Medicine in Bogotá, Editorial Minerva, 1935, 17.

37. *Ibid.*, 40.

38. Christopher Abel, *External philanthropy and domestic change in Colombian health care*, 339.

39. Letter from Juan Guiteras to W.C. Gorgas, August 28, 1916, RAC, RFA, RG 5, Series 2, sub-series 311, box 27, folder 163.

40. In this investigation, the diagnosis of yellow fever relied heavily on laboratory tests both in Bucaramanga and in Bogotá, using both in vitro reactions with *Leptospira icteroides* and innoculation of guinea pigs. Although contended, and later to be demonstrated false, *Leptospira icteroides* was at that time believed to be the etiological agent of yellow fever. The Rockefeller Foundation carried out a series of works and vaccinations based on this belief. See Cueto, "Sanitation from above."

41. W. A. Sawyer, Public Health in Colombia. RFA, RG 5, Series 2, Subseries 311, Box 27, folder 162.

42. Cueto, "Sanitation from above."

43. *Revista de Higiene*, Vol. 11, nº 144-145, enero de 1924, 44.

44. Steven C. Williams, "Nationalism and Public Health: the Convergence of Rockefeller Foundation Technique and Brazilian Federal Authority during the Time of Yellow Fever, 1925-1930," in Cueto, *Missionaries of Science*, 34.

45. R. Franco, J. Martinez Santamaria, and G. Toro Villa, "Fiebre amarilla y fiebre espiroquetal," *Academia Nacional de Medicina, Sesiones científicas del Centenario*, (Bogotá, 1911).

46. Fred L Soper. "Rural and Jungle Yellow Fever: a New Public Health Problem in Colombia," Lecture given before the Faculty of Medicine of Bogotá, April 5th, 1935. Bogotá, Editorial Minerva, 1935, 17.

47. Soper, *Rural and Jungle Yellow Fever*, 20; J. A Kerr and Lius Patino Camargo, "Investigaciones sobre fiebre amarilla en Muzo y Santander," *Revista de Higiene* (Segunda Epoca) 2.3 (1933): 63-91.

48. D. B. Wilson to Sawyer, March 10, 1936, RAC, RFA, RG 1.1, Series 311, subseries Colombia, box 2, folder 11.

49. Christopher Abel, "External Philanthropy and Domestic Change in Colombian Health Care," 361.

50. Malariologia, *Revista de Higiene* 23,4-5(1942): 10.

51. George Bevier to H. H. Howard, July 24, 1934. RAC, RFA, Series 311, box 7, folder 73.

52. *Ibid.*

53. Catherine Legrande, *Frontier Expansion and Peasant Protest in Colombia, 1850-1936* (Albuquerque: University of New Mexico Press, 1986), 98.

54. Colombia Departamento Nacional de Higiene, *Unidades Sanitarias y Comisiones Rurales*, (Bogotá, 1936).

55. Christopher Abel, *Health Care in Colombia c.1920-c.1950: A Preliminary Analysis* (London: Institute of Latin American Studies, 1994), 42.

56. Dr. J. H Paul, Journal, February 17, 1936. RAC, RG 1.1, Series 311, Box 12, Folder 108.

57. H. H. Howard to Dr. G. Bevier, August 15, 1933. RAC, RG 1.1, Series 311, Box 10, Folder 92.

58. George Bevier to Enrique Enciso, September 21, 1933. RAC, RG 1.1, Series 311, Box 10, Folder 92.

59. George Bevier to H.H. Howard, September 26, 1933. RAC, RG 1.1, Series 311, Box 10, Folder 92.

60. George Bevier, Annual Report for the Year 1934 on the Cooperative Health Activities in Colombia. RAC, RG 5, IHB/D, Series 3, Subseries Reports, Routine, Box 132, Folder 1557.

61. Fred L. Soper to F. F. Russell, February 23, 1935. RAC, RG 1.1, Series 311, Box 10, Folder 94.

62. Fred L. Soper to the Director of Hygiene, March 25, 1935. RAC, RG 1.1, Series 311, Box 10, Folder 94.

63. Lewis Hackett, LWH Notes for IHD history vol. III, Pages 991-1544. Interview with Dr. Hugh H. Smith, June 22, 1951. RAC, RG 3, Administration, program and policy, Series 908, Box 4, Folder 21.

64. Journal of Austin Kerr, August 20, 1936. RAC, RG 1.1, Series 311, Box 13, Folder 109.

65. Journal of J. A. Kerr, October 13, 1936. RAC, RG 1.1, Series 311, Box 13, Folder 109.

66. Lewis Hackett, LWH Notes for IHD history vol. III, Pages 991-1544. Interview with Dr. Hugh H. Smith, June 22, 1951. RAC, RG 3, Administration, program and policy, Series 908, Box 4, Folder 21.

67. Austin Kerr Diary, July 9, 1937. RAC, RG 1.1, Series 311, Box 13, Folder 109.

68. Austin Kerr Diary, March 3, 1938. RAC, RG 1.1, Series 311, Box 13, Folder 109.

69. Letter from W.A. Sawyer to F. L. Soper, August 4, 1937. RAC, RG 1.1, Series 311, Box 11, Folder 98.

70. Letter from F. L. Soper to W. A. Sawyer, May 11, 1938. RAC, RG 1.1, Series 311, Box 11, Folder 99.

71. W. A. Sawyer to F. L. Soper, May 23, 1938. RAC, RG 1.1, Series 311, Box 11, Folder 99.

72. Academia Nacional de Medicina to R. B. Hill, May 19 1946. RAC, RG 1.1, Series 311, Box 11, Folder 104.

73. SMITH, Hugh H., Diary. April 30, 1939. RAC, RFA, RG 1.1, Series 311, Box 13, Folder 113.

74. Marston Bates, Diary, December 11, 1941. RAC, RFA, R.G. 1.1., Series 311, Box 13, Folder 113.

75. Additional designation and revision of budget, June 6, 1946. RAC, RFA, R.G. 1.1., Series 311, Box 11, Folder 101.

76. Nancy Bell Bates, *East of the Andes and West of Nowhere: A naturalist's wife in Colombia* (New York: Charles Scribner's sons, 1947), 17.

77. R. B. Hill to P. J. Crawford, September 21, 1946. RAC, RG 1.1, Series 311, Box 11, Folder 104.

78. R. B. Hill Diary, January 19, 1946. RAC, RG 1.1, Series 311, Box 14, Folder 115B.

79. R.B. Hill Diary, January 24, 1946. RAC, RG 1.1, Series 311, Box 14, Folder 115B.

80. Bates to Strode, January 18, 1948. RAC, RG 1.3, Series 311, Box 8, Folder 77.

81. J. Elmendorf to R. B. Hill, January 12, 1948. RAC, RG 1.1, Series 311, Box 8, Folder 77.

82. J. Elmendorf to R. B. Hill, January 15, 1948. RAC, RG 1.1, Series 311, Box 8, Folder 77.

83. N. Miranda, E. Quevedo, and M. Hernandez, "Medicina (2): la institucionalización de la medicina en Colombia," in *Historia Social de la Ciencia en Colombia, Tomo VIII* (Bogotá: Colciencias, 1993), 136. Augusto Gast Galvis, *Historia de la Fiebre Amarilla en Colombia* (Bogotá: Instituto Nacional de Salud, 1982), 26.

84. Data taken from Table 33, International Health Division of the Rockefeller Foundation Yellow Fever Expenditures, 1916-1949, in Strode, "Costs and Man Power" *Yellow Fever*, 637.

85. L. Hackett, LWH Notes for IHD history vol. III, Pages 991-1544. Interview with Dr. Hugh H. Smith, June 22, 1951. RAC, RG 3, Administration, program and policy, Series 908, Box 4, Folder 21.

Chapter Seven

"Working through government" and "working in close co-operation with government": A Perspective on the Rockefeller Public Health Program in Colonial India

Shirish N. Kavadi

In September 1952, Robert S. Morison, Director, Division of Medicine and Public Health (DMPH) in the Rockefeller Foundation (RF), then on a visit to Lebanon recorded in his diary that given the corrupt nature of the government there, it had become exceedingly difficult to work with official agencies in that country. Instead, he suggested that the DMPH work with voluntary or non-official agencies, not just in Lebanon but the entire Middle East and even parts of Latin America. Morison underlined that his views applied as much to governments in general and were not restricted to corrupt governments or rigid bureaucracies. His opinion received endorsement from Dr. E.T. Mehra of the Iranian Foundation. Recounting his meeting with Mehra, Morison noted that Mehra attributed the RF's "failure" in Iran to what he said was its insistence on "working through the government" rather than "working in close co-operation with government." In addition, according to Mehra, the Iranian government had "failed completely to grasp the attitude, partly because it was not familiar with the procedure, and partly because it found it difficult to believe that the RF did not have as much money as the name might imply." He further believed that "it would have been much better if the RF project had been organized quite separate from government agencies."[1] Morison's observations and Mehra's comments highlight the uneasy relationship that has existed between philanthropic foundations and governments and the problem areas that very often afflict their association.

Since the early 20th century, philanthropic foundations have played a principal role in the development and expansion of scientific medicine and public health. Among philanthropic organizations, the RF played a pioneering role in globally extending the influence of scientific medicine and modern public health practices. For the RF, spurred by the ideal of working for the "well being of mankind," scientific medicine and public health were the primary instruments to meet this goal. The RF conceptualized a model of public health based on its American experience, which it sought to propagate worldwide. RF officials had great faith in their approach and zealously promoted it.[2] Occasionally working independent of governments, foundations have mostly sought the latter's co-operation and endorsement. The RF, for instance, as a matter of principle and policy, collaborated only with governments. Underlying this move was a certain assumption: that such collaboration would ensure accountability, stability, acceptance, and most importantly results. In other words, the successful outcome of a specific program would depend upon ensuring government co-operation. Governments on their part have turned to philanthropic organizations seeking technical advice, as well as personnel and financial resources in the furtherance of medicine and public health. Notwithstanding an element of mutual interest and reciprocity, it has not been uncommon to find that ambiguity, unrealistic expectations and dissatisfaction have characterized their relationship. This eventually has a bearing on the outcome of the program.

The distinction Mehra made between "working in close co-operation with government" and "working through government" is significant for it throws light on the nature of relationships between foundations and governments. In making the distinction, Mehra was suggesting that there are different ways in which foundations approach or may approach governments. Moreover, he was indicating that one approach ("working in close co-operation with government") was more appropriate than the second ("working through government") or the preferred one. The first suggested partnership and an equal relationship. The second implied a forcible intervention or imposition, and the narrow pursuit of personal goals. Within these approaches, he sought to locate the issue of "success" and "failure" of the Foundation's program. It is doubtful whether the differentiation made by Mehra existed in the minds of RF officials. In the zealous pursuit of their global goals, there was little time for such fine distinctions. Nevertheless, is it possible to make such clear differentiation? Governments are not single entities. With their complex structure and many tiers there is no one single worldview or shared perception. Foundations have to work with each of these levels and authorities and how each respond would differ upon their own perceptions and priorities. Mehra's remarks about the failure of target governments to grasp RF principles or the misconceptions regarding the availability of funds from the Foundation are common and familiar. The problem at least partly arises because of this complex decision making process. Probably this was to be expected. Sometimes, for the officials of the RF, getting the government to accept a program outweighed all other objectives giving cause to these misconceptions among governments. It also led to deviations from the original scheme and the compromising of important principles.

What these observations and remarks speak of are differing perceptions, priorities, expectations and attitudes all of which have bearing on the nature of collaboration between foundations and governments. Negotiations between the two bring out the differences in perceptions and priorities and reveal those that are accommodated. The purpose of the present chapter is to examine the manner in which foundation officials and co-operating government agencies negotiate the undertaking of a program, and the subsequent interplay of different interests, concerns, and perceptions. Is it possible to discern in these negotiations the nature of the approach and the course the relationship would take? The specific context is the RF's negotiation for a demonstration rural health unit with the Madras provincial government in colonial India.

The Rockefeller Foundation and Public Health in India

The RF public health program in colonial India began against the backdrop of the devolution of power to the provincial governments. This was in part the outcome of the Central Government's economy drive and in part as a response to nationalist demands for greater role in government. Provision of health care and education became the responsibility of the provincial governments. The provinces were beginning to undertake the first tentative steps towards organizing public health departments and activities. The RF's offer to conduct various co-operative public health activities that included disease control campaigns, establishment of rural health units and training of public health personnel seemed timely. RF public health programs aimed at demonstrating effective, efficient and economical methods of disease control laying stress on public health education and community participation. The RF's work did not span the whole of India but was restricted only to those provinces or Princely states that showed a willingness to co-operate with it.[3] Among the first to collaborate with the RF was the Government of the Madras Presidency in southern India. Before we take up the discussion of the RF's efforts at developing rural health units in Madras a recapitulation of the public health situation in British India and Madras in particular will be in order.

Public Health in India

The RF's intervention in Indian public health began immediately after the end of the First World War. It happened at a juncture when public health in India was witnessing a transition. Western medicine was introduced into India as early as the fifteenth century, but scientific medicine and modern public health had its origins in nineteenth century British colonial policy. Medicine under colonial rule had limited scope. It was essentially curative and its coverage was primarily aimed at the army and European enclaves. It neglected preventive care for Indians, particularly in the rural areas, and therefore was barely effective in ameliorating the health conditions of the vast majority of subject population.[4] The post 1919 years saw Western medicine partly breaking out of its enclavism.[5] It was not a radical transformation of the colonial health policy but a reorientation of state policy away from the enclavism of the past. Medical research and sanitary practice moved more towards an Indian oriented system of public health.[6] This change in the government's approach towards public health was as much a result of advancement in scientific medicine as it was due to economic factors and political developments in India.

The 1890s had inaugurated a new phase in the political and constitutional evolution of India with significant consequences for the administration of public health. The devolution of power to local political bodies had begun as early as the 1880s. Political rule had passed from the British East India Company to the British Crown in 1858. The consolidation of its political control necessitated the expansion of its military presence and the creation of an administrative structure. This led to an increase in military and administrative costs. To prune the operational costs of running the country, changes were undertaken to the administrative structure through decentralization. Baker[7] says the dominating factor behind these administrative changes was the lack of money. In the process, as is common to most government efforts at curtailing expenditure, non-productive sectors such as social services suffered. Government claimed increasing difficulty in finding funds to provide social services and to develop India's economic resources. The government's response to the problem of expanding revenues and administrative responsibilities was successive devolution of power. Provincial governments offloaded to lower bodies responsibility for public health and sanitation along with other social sector services such as education. The First World War further weakened British India economically. The War turned the existing financial stringency into a bigger financial crisis with the government having to face budget deficits and debts.

The gathering momentum of the nationalist movement led by the Indian National Congress (INC) pressured the colonial government into granting constitutional concessions to the Indians. There was also demand from the provinces for more autonomy to adjust their internal affairs to the special circumstances of their particular regions. The most significant fallout of both the financial crisis and demand for constitutional reforms in the early decades of twentieth century was the introduction of Dyarchy under the Government of India Act of 1919. The Act made a firm division between central heads of revenue and those of the provinces and granted a limited autonomy to the provinces. Indians were given representation in the Legislative Council through elections. The Governor's Executive Council now consisted of official European Members and a few elected Indian representatives co-opted as ministers. The devolution actually consisted of transferring administrative burden and not executive power and the presence of responsible Indian ministers in the Legislative Council was to counter discontent among Indian elites. Sanitation, vital statistics and public health was transferred to provincial control. Indian ministers were allotted charge of such "soft" portfolios as education and health. Official European Members of the Governor's Council retained control over finance. With control over finances remaining with European members of the Governor's Council, any proposal submitted by Indian ministers faced budget cuts. This, in short, was the nature and functioning of Dyarchy.

Among the three Presidencies of Bengal, Bombay and Madras, the last one was not a particularly rich province. Nevertheless, right from the 1870s and into the 1930s the Government of India depended upon Madras to meet a large degree of its financial requirements. Whenever the other two Presidencies willfully or otherwise failed to pay up even a part of their share to the central treasury the Government of India would put pressure on the Madras Government to make up for the deficit. In 1919, a British minister in the House of Commons described Madras as a "milch cow." During the depression years, to balance his budget the Madras Finance Minister decided to trim government expenditure. Instead of being rewarded with a generous and increased annual central grant, the annual Central Government reduced its contribution to Madras.[8] Without denying the overall effect of this on government expenditure, it must be pointed out that this factor also became an excuse for reducing government expenditure on such services as sanitation and irrigation. As a result, the RF's intervention was welcomed because of the funding that was on offer for certain public health measures.

The transfer of public health to the provinces required provincial governments to undertake measures to develop public health. Existing public health systems were ineffective, inefficient and inadequate while awareness about public health issues was growing. Provincial governments initiated various measures to develop a broad health care delivery system. For instance, after 1920 governments attempted to find common ground between indigenous medicine and Western medicine. Personnel for rural services were trained with efforts directed at using local people as the bottom tier of the medical and sanitary services. Governments introduced schemes for village nurses and for schoolmasters as health personnel.[9] The Madras Presidency, for instance, set up a Board of Public Health in 1920 though it remained inactive.[10] A more important development in health care in Madras was the introduction in 1924 of a scheme specifically formulated to provide health care to the rural population. Called the Subsidized Rural Medical Relief Scheme, it essentially consisted of encouraging private medical practitioners to settle down in villages by offering them a monetary incentive in the form of a subsidy. Most importantly, governments decided to organize public health departments on sound administrative principles to make them effective and efficient.

The RF Public Health Program in Madras: Poonamallee Health Unit, 1935-1940

The RF's earliest public health intervention, an anti-hookworm campaign, was centered in the Madras Presidency.[11] Indentured labor sent from the province to the British colonies had been identified as the source of hookworm infection amongst the population of these colonies. The infection rate was as high as ninety per cent. The RF believed that the disease had to be controlled at its source. The campaign, apart from aiming at educating people in hookworm control, was to be an entering wedge for extensive sanitary programs. Most importantly, it was meant to create "a health concept in the minds of people" and "a desire for public health agencies." In popular and medical perceptions hookworm was not yet regarded as a major public health menace. Despite large-scale prevalence of the hookworm infection, the British Indian Government, which did not view it as a major cause of disease and a threat to public health, had shown a singular lack of concern for undertaking organized eradication or control measures.[12] John Ettling[13] observes that before the RF could undertake hookworm control, it had to spend energy and money to create the disease in the minds of people in whose perceptions it was not an illness and a threat to individual health but an everyday fact of life like the common cold. Investigations conducted before and during the course of the campaign revealed that though infestation was high it did not have a very detrimental effect on the health of the people. This may be explained by the fact that the most common species found in the region was the harmless *Necator americanus* species that affected nearly 100 percent of the population in most areas. *Ancylostoma duodenale*, the more pernicious of the two commonly known species, was found affecting a much lower percentage of the populations and in only a few localities. A great deal of ambivalence characterized the official stance on hookworm control. Government and tea estate officials oscillated between apparent enthusiasms for the campaign to indifference towards actual implementation, evident in their reluctance to spend on the construction of latrines. The outcome of the campaign was mixed: mass treatment became a routine practice in government health institutions but sanitation, which was crucial to controlling hookworm infection, was neglected. The infection rate in the province failed to decline.

Limited success or even failure of a particular program was no deterrent for the RF in its pursuit of the "well being of humankind." Even as the hookworm campaign was coming to an end RF officials were planning the next phase of their public health program in the Presidency. In the early 1930s, the RF had begun exploring the possibilities of setting up demonstration rural health units in India. Health units, to be established at a micro-level in rural areas, were an extension of the public health work that had begun with the hookworm campaign. Through the medium of the health units the RF proposed to demonstrate feasible administrative methods for public health organization, economical means of running rural public health work, and techniques of health education and community participation. Eventually these would evolve into training centers for public health work. Provincial governments and governments of princely states were approached. Many Directors of Public Health in the provinces expressed interest in the demonstration health units.[14] Support for the health units also came from Maj. General Megaw, Director General of the Indian Medical Service.

The RF offered assistance to participating governments in setting up the health unit. Apart from exposing the basic principles underlying the concept of the health and defining the objectives and activities that the health units would conduct, the RF would offer technical advice and part of the required finances. In addition, the RF offered public health training fellowships to medical officers nominated by the Government to prepare them to assume responsibility for health unit activities.

In the conduct of health unit work, Dr W. P. Jacocks, the RF representative who introduced and developed health units in Ceylon and India, stressed the need for co-operation between those responsible for running the health unit on the one hand and the community and government on the other. The key word was co-operation. Integral to it was the acceptance of the universality of public health principles, the absence of compulsion and imposition, and the assurance of active community participation. Jacocks spelled this out observing, "Public health work done by compulsion has no sound foundation." While admitting that the "health unit scheme presupposes that fundamental public health principles are universally applicable," Jacocks recognized that local conditions had to be given consideration while determining procedures to be followed in health unit work. These, he said, would have to be based on common sense and "on a recognition of what is useful reasonable and practicable." RF representatives claimed that the proposed health units were not a direct transplant of the health units evolved in the USA. Jacocks argued that all health units in Asia were adapted to the local conditions and only then put into practice. The single most important consideration was that "a unit should not be established in any community which does not evince an active interest in the subject and show a desire to co-operate in every way."[15]

The negotiation between the RF and the Government of Madras focused on two main issues: 1) the period of co-operation; 2) RF's financial contribution. The time factor involved a basic principle—that a definite minimum period was essential to demonstrate the efficacy of certain public health methods and practices. A tenure determined in advance would also ensure that the amount of RF funding was restricted. It was important to convey to the recipients that Rockefeller largess was not unlimited. The RF, during the hookworm campaign, had considered five years as adequate to demonstrate control methods. When the proposal for the health unit was initiated, the RF had yet to take an official position on the duration of co-operation and hence there were no guidelines. Soon after the initial proposal was submitted, the RF decided on five years as the appropriate period of co-operation. The health unit set up at Poonamallee in Madras ran for a period of five and half years as against these five-years. At the bidding of the Government of Madras, the health unit was inaugurated on an experimental basis for a limited term of one and half years and was later extended by four years. This was an important digression from the established pattern. Local political and financial considerations were important factors in influencing the decision regarding the period of demonstration. Financial share was a major issue between the two agencies. The annual budget for a unit was estimated at Rs.30,000/-. The RF's share was to be forty percent of the total expenditure which it provided through an agreed scale of payment, the proportion diminishing each year until after the fifth year when the entire financial responsibility would pass on to the co-operating Government. The Government hoped to get the Foundation to agree to enhance its share while its own share was very often not committed.

The RF, as mentioned earlier, had already started exploring the possibilities of creating a health unit, but its hopes of setting up one in Madras were raised when in 1934 Lt. Col. Dr. C. M. Ganapathy, IMS was appointed Director of Public Health.[16] Ganapathy was the first Indian IMS officer to be elevated to the post of Director in the Presidency. Dr W. P. Jacocks, the RF representative stationed in Madras at the time, "persuaded" Ganapathy of the usefulness of a rural health unit. Ganapathy responded with a proposal that he submitted to the government in December 1934 in which he recommended a three-year period for co-operation and the RF financial contribution at a decreasing scale of one half, one third, and one quarter for each of the three years. The Scientific Directors of the IHB, as stated above, had yet to decide upon the official and appropriate co-operation period for health units and hence, at this stage, the RF offered no fixed guidelines regarding the duration. However, soon after in early 1935, the Scientific Directors gave their approval for a five years period of co-operation. The RF also determined their own contribution at 60, 50, 40, 30 and 20 percent respectively for each year of the period of co-operation. Ganapathy revised his original proposal according to the guidelines provided by the Scientific Directors and submitted the same to the government.

A great deal of uncertainty hung over the proposal as the Government appeared unwilling to act upon it and give its sanction. The Government's silence was a reaction to the existing political situation in the Presidency. In the 1930 elections held to the Legislative Assembly in Madras Presidency, the non-Brahmin Justice Party won and some of its elected representatives appointed ministers in the newly formed Government.[17] Muniswami Naidu of the Justice Party was made the Premier but in 1932, the Governor of the Province Lord Erskine replaced him with the Raja of Bobbili. The Raja, who had to be commanded and goaded into taking any action, did not enjoy the full confidence of the Governor and the British service officials. On the other hand, viewed as a "yes-man" in non-official circles, the Raja had no control over the party. His position within the party further weakened as factionalism grew within the Justice Party and the Party began to crumble. With its main rival the Indian National Congress, seen as a party of Brahmins and upper castes, on the ascendance defections from the Justice Party to the Congress debilitated it further. In the elections to local bodies between 1934 and 1936 the Congress gained control over one third of the local government bodies in the provinces. A consequence of all this was that the Justice Party Ministry failed to pass any important legislation.

In the early 1930s, the British began to disengage themselves further from the Raj. As a part of this process, it was decided to give the provinces more autonomy. A major constitutional step in this direction was the Government of India Act of 1935, which abolished dyarchy and introduced provincial autonomy. The changes in provincial government may have appeared as an important constitutional victory for the Indian nationalists but the British ensured that they retained strong and rigid bureaucratic machinery in the provinces.[18] Notwithstanding the political and constitutional developments, or rather more because of it, the bureaucracy remained committed to the values of the Raj.

The changing political scenario with new constitutional reforms in the offing caused apprehensions within the Government about its own political survival and the likelihood of its being re-elected under the new Constitution. It was this uncertainty that caused the Government to withhold making any decision. The delay was to cause anxious moments among RF officials and the Director of Public Health. It appeared that Dr. W.C. Sweet, the RF field representative based in Bangalore (in the princely state of Mysore in south India) and Ganapathy would have to work hard to get the Government to act and sanction the project. They found the attitude of the Health Secretary in the Government unhelpful and his dilatory tactics most pronounced. After a great deal of argument and a good deal of bargaining over contributions, Ganapathy and Sweet were able to persuade the Secretary to accept the scheme. Despite convincing the Secretary the Premier, the Raja of Bobbili demonstrated the same unhelpful attitude and unwillingness to commit the Government to the scheme.[19] Apparently, the Government position was that it considered it inappropriate to commit any incoming Government elected to power under the new Constitution to any long-term agreements.

The attitude of the Raja may have been partly determined by his diffident personality and the existing political situation but the fact was that Indian ministers had a limited say in decision-making. It is more than probable that the Secretary and the Finance Member—both Europeans—prevailed over the Minister, who was not just perceived as irresolute, but was actually quite pliant. The Government continued to waver. Given this stand of the existing Government Dr. Jacocks, the RF India representative based in Delhi, considered it prudent not to pursue the proposal until the new Indian Constitution became operational in April 1937 and a new Government was formed.[20]

Notwithstanding its somewhat dubious stance, the Government did not ignore the proposal altogether. In April 1935, the Government informed the Director of Public Health that they considered the five-year period too long and revived the three-year proposal except that the suggested RF contribution was to be in the proportion of 60, 50 and 40 percent respectively for each of the years.[21] Sweet considered this unacceptable and instead offered to the Government the two original options: a three year co-operative period at one half, one third, and one quarter, or a five year period at 60, 50, 40, 30 and 20 percent. Sweet strongly recommended the latter but personally was not averse to the former.

The Government reiterated the claim that its sense of propriety did not permit it to commit any incoming government to a decision taken by it. Simultaneously unwilling to give up the project, the Government came up with an altogether new alternative. The Government offered to undertake health unit work on an experimental measure for a period of eighteen months beginning 1 October 1935 to end on 31 March 1937.[22] Suggesting an alternate plan with shorter duration was the Government's way of ensuring that any new Government was not bound by its decision. With regard to its own financial share, the Government hesitated to make any commitment and stated, "The question of further continuance of the share will depend on the availability of funds and the vote of the Legislative Council."[23]

The Government's new proposal was purely experimental, a departure from the IHD's accepted procedure and outside any of its usual agreements. Consequently, Sweet's first thought was to refuse this co-operation but his pragmatic side appears to have dictated adjustment with the Government. He justified this readiness to accommodate on the grounds that, "in view of the enthusiasm of the Director of Health, the possibility that this would be about the only way to get Government's consent to any health scheme," and the fact that the Government order held out the possibility of continuance of the unit after the eighteen months were over. Sweet recommended to Jacocks that the RF accept the new plan and co-operate. To resolve the question of the RF's contribution to the project during this period, Sweet suggested that the RF pay the average percentage of total contribution for a five-year period. This worked out to 40 per cent for eighteen months co-operation. Jacocks, too, was skeptical but eager to get early sanction for the project. Hence, overcoming his scepticism, he accepted Sweet's plan.[24]

Although Jacocks gave into Sweet's urging for co-operation, he did not conceal his thoughts. In a letter to Col. Ganapathy he candidly expressed his reservations but indicated that the commencement of the scheme was more important. Jacocks observed,

> We both knew of course, that no satisfactory results can be shown by any health organisation in eighteen months. For this reason, I am not particularly interested in a co-operative plan for that period of time and under ordinary circumstances would not pursue the subject further. At the same time I well remember from conversations with you, your earnest and sincere desire to improve rural health methods in the Madras Presidency and your conviction that a health unit would be of great assistance in your plans. With this idea in mind I write to say that if you desire to undertake this sound work and think that in eighteen months its efficiency can be partly demonstrated, I am willing to ask our organisation to co-operate on the basis of paying 40% of the budget (Rs30,000/-) for a period of eighteen months. If later it is desired to extend the work for the usual five year period, consideration will be given to adjusting our co-operative contributions to the usual five year basis.[25]

With the Government and the RF agreeing to an experimental period of eighteen months with a 40 percent contribution from IHD, the Government gave its sanction in August 1935[26] and the Scientific Directors approved the scheme the following month.

The health unit was inaugurated in Poonamallee in October 1935. The scheme, as the Director of Public Health explained, was to be conducted in co-operation with the community and the Panchayat Board in the area. It was intended to try intensive health work as an experimental measure in one selected area "with a view to demonstrate what could be done for the health and welfare of the rural population by a well organized public health effort and thus to stimulate the health consciousness of the masses."[27] The scope of the new scheme was wider for its coverage included the vast, neglected, rural population.

The Revised Scheme for Poonamallee

The commencement of the health unit on an experimental basis was only a temporary arrangement. The issue of its continuance after March 1937 and eventual conversion into a regular full-fledged demonstration health unit remained to be decided. The matter came up for discussion between Ganapathy and Sweet. There appeared two possibilities: 1) the Government would run the unit until 31 March 1937 which would give the incoming Government time to make up its mind on continuation; 2) the unit should be continued for four years from the 1st of April 1937 making five and a half years the total period of co-operation. Sweet made a generous offer if the Government accepted the shorter period of co-operation. If the Government did so, he offered to have the RF continue to contribute 40% of the expenses until 31 December 1937 so that the new government would have the time to consider its decision. If the second proposal with four years co-operation were accepted, the RF would contribute 50 per cent in the first year and in subsequent years 40, 30, and 20 per cent. These figures were arrived at to carry out Jacock's suggestions that the RF would adjust its own co-operative contributions to the usual five-year basis. According to Sweet this distribution of RF contributions, with the previous 40 per cent contribution for the eighteen month trial period, added up to the 40 per cent contribution over five years to other health units.[28]

Jacocks, no less keen than Sweet to see the project through, approved of the latter's suggestions. He agreed that the Government's continued co-operation until 31 December 1937 would "give the Government an opportunity to consider what its future policy would be." Getting the Government to co-operate for the full five-year period was the objective. Jacocks further agreed to co-operation based on 40 per cent contribution for as he observed, "When the co-operation period is extended to five years, we can adjust the proportions so that eventually the Government will receive from us what it would have received if the usual co-operative programme had been put into practice."[29]

The appointment of a new Secretary to the Government appeared to have raised a new difficulty. The Secretary, though sympathetic to the scheme, had some reservations regarding its practicality; it was with some reluctance that he consented to forward the proposed plan to the Government. Finance was a problem but the Government decided to avail funds from a grant made to the provinces by the Government of India. Without much difficulty, the Madras Government managed to get the Government of India's approval which was essential. The Madras Government issued an order on 2 August 1936 approving the continuation of the unit for a period of four years from 1 April 1937. For Jacocks and Sweet the matter did not end there. They had succeeded, with less than expected difficulties, in getting the necessary approvals from the Madras

Government for a five-year co-operation for the health unit. Convincing officials in the New York office was the next major task.

On 11 August 1936, Jacocks conveyed to Wilbur Sawyer in the New York office the budgets for the co-operative period and the proposals for a four-year period of co-operation at the rate of 50, 40, 30 and 20 percent. The proposal did not find acceptance among the officials at the RF. The changes incorporated to accommodate the Madras Government were viewed as a deviation from regular RF practice and officials in New York were not entirely convinced this would help them to further their program. Sawyer, for instance, stated very clearly that he did not consider raising the percentage to be good practice and that he would prefer continuing October 1st as the yearly date. He asked Jacocks to revise the proposal and to regularize it to the extent possible. Sawyer argued that the proposal was essentially a continuation of the existing project toward which the IHD would have been anyway paying 40 per cent of the budget for a period of eighteen months beginning 1 April 1937. As far as the date of commencement was concerned he believed that unless there was a real advantage in having the year begin on April 1, he would prefer to have the project continue on the schedule on which it was started, which was the year beginning October 1. Adhering to the original schedule, he pointed out, would leave the RF with three and one half years more instead of the suggested four. He considered the most serious irregularity in the proposal to be the increase in the proportion of the Foundation's contribution in the middle of the second year of the project from 40% to 50% of the budget. RF officials were not inclined to recommend this increase. Sawyer instead directed Jacocks to ascertain whether Government desired RF co-operation based on a continuation of the 40% proportion through the third year and the proposed 30% and 20% in the fourth and fifth year. Sawyer's essential argument was that the irregularity in the budget and the deviation from normal practice appeared to have no rationale and justification behind it.[30]

Sweet and Jacocks apprised the Director of Public Health of Sawyer's views. Ganapathy conceded that the change of the closing date would make little difference to the Government and so accepted Sawyer's condition. However, in the matter regarding financial contribution the situation was different and more difficult. Sweet pointed out that the Government had been persuaded to agree to the Poonamallee Health Unit scheme with great difficulty. Any change in proportions worked out by the IHD would necessitate the reopening of the entire issue with the Central Government. Ganapathy feared that this would probably result in the Government of Madras refusing to continue with the matter. Sweet shared the anxiety of the Director and urged Jacocks to request the IHD to give approval to the earlier proposition.[31]

Sweet proposed a period of three and one half years with the RF share at 50, 40, 30 and 20 per cent for the last six months.[32] This proposal by Sweet also did not find acceptance with Sawyer, who raised certain important policy matters. In a letter to Jacocks, Sawyer observed:

> We would not wish to enter upon the projects unless they were planned for the full period. . . . I have read the explanation of the increase in our percentage in the middle of the Poonamallee health unit project, but I am not entirely satisfied with it, as the increase in the middle of the project seems illogical unless there are additional reasons for the increased liberality. In this project, as in the others, we assume that Government will take over all the costs at the end of the five years so that we complete the project and enter into new fields.[33]

Jacocks responded stating that Sweet's and his actions were dictated by a desire to help Col. Ganapathy to get the health unit well established for five years and to secure the full amount of financial co-operation from the RF that he would have received had he gone for the regular scheme. He pointed out that while Ganapathy was enthusiastic about the health unit his problems were with the Government. Since he was the RF's hope for the unit, it was necessary to help him out. Jacocks pointed out, "it is my impression that the Madras Presidency has always proved most difficult to deal with and that may explain why the Doctor finds trouble in getting a decision from them."[34] Government then was a major problem and the earnestness of an individual official was not enough in developing co-operation.

Despite reservations, the RF showed enough flexibility in its approach. It was not just willing to accommodate Ganapathy but even ready to compromise on certain fundamental principles. Keen that the health unit serve its full term of five years and demonstrate its utility and efficacy, the RF went ahead in committing its share of the finances. In a meeting held on 29 October 1936, the Scientific Director through a resolution approved an additional designation of Rs.39,000/- for the three and half year period from 1 April 1937 to 30 September 1940 on a descending scale. The Government on the other hand caused much disappointment and dissatisfaction. Its failure to grasp RF principles was evident in its parsimonious behaviour towards its own financial commitments. In the functioning of the health unit, this failure was reflected in the working of the staff. In Feb 1937 Sweet wrote to Jacocks to express his unhappiness over the Government keeping the health unit budget below the required Rs 30,000/-. The Government was disinclined to disburse its share of the funding. This was, according to him, a hindrance to its work. Sweet remarked, "There seems to be a general attitude that the work should be made as cheap as possible." The attitude of government health workers was a greater difficulty, for as he noted, "possibly the most unsatisfactory part of the Poonamallee work is found in the efforts of the sanitary inspectors who retain a considerable amount of official, law-enforcing attitude in their areas."[35] None of this, however, made the RF to reconsider its relationship with Madras or suggest discontinuation of the program. If anything, continuation of the program for the full five years under any circumstances seemed to be the concern.

Elections under the 1935 constitutional reforms were held in February 1937 and provincial autonomy was introduced with effect from 1 April 1937. The elections also brought the Indian National Congress into power in Madras and in several other provinces of India. A brief interlude of political uncertainty followed. In July 1937, a new Ministry under Mr. C. Rajgopalachariar as the Chief Minister was sworn in. It had Dr. T.S.S. Rajan as the Minister for Public Health. In October 1939, the Congress Government in Madras, along with Congress Ministries in other provinces, resigned in protest against the British Government's decision to involve India in the war against Germany. The formation of a Government by the Congress appears to have made little difference to policy. Indeed, the Congress Government displayed an even more baffling attitude. Before it laid down office the Congress Government presented a memorandum recommending that at the expiration of the co-operative period the work should be stopped and a "better but cheaper arrangement should be made." Madras came under the direct rule of Governor Sir Arthur Hope who was assisted by British officials from the Indian Civil Service and a few handpicked Indians. In February 1940, the Director of Public Health was able to successfully persuade the new Government to reconsider the recommendation at least partly because there was local demand for continuation of work after September 1940.[36] With no more money coming from the RF the Madras Government was unwilling to allocate the full cost required for running the health unit on the scale planned by the RF. The Government decided that it would continue the Poonamallee health unit at its own expense "on a modified scale." This was to Jacocks "an improvement on the first intention by the Congress Government to abolish it."[37] After September, the health unit became an activity of the Madras Public Health Department functioning as a field-training center for medical officers who were to be posted in rural areas. The Government even appeared well disposed to the idea of making the health unit a permanent Government program.[38]

The health unit being run on a "modified scale" faced several inadequacies and shortcomings. Sometime in 1946 when Dr. M. C. Balfour, the RF representative in India, visited the unit he noted the inadequacies and deficiencies in its working and suggested the appointment of a first class Health Officer for adequate supervision and administration: Major J. H. Gorman, IMS, the Director of Public Health. Madras indicated this would be possible if the RF could meet the salary and allowances of this officer. Balfour turned down the idea.[39] In April 1946, Gorman requested RF financial aid for the Poonamallee health unit. Once again, Balfour declined stating that the question of the RF's participation was closed. Moreover, he added that both Dr. George Strode from the New York office and he were of the opinion that the original plan was not likely to succeed in establishing a satisfactory training and research center on the lines conceived. Balfour claimed that they had arrived at this conclusion after assessing the scheme and the local situation.[40]

The health unit had deviated from the original scheme and the RF had little expectation that the scheme could be brought back on track. The RF managed to persuade the government to set up the health unit. It pushed the project through overcoming various obstacles but failed to convey the underlying principles. The various modifications the RF had to make to the scheme required setting aside regular RF practice. Differing perceptions and varied degree of interest at different tiers of the government made it that much more difficult for the RF to work in close co-operation with the Government. In its eagerness to see the scheme materialize, the RF had to perforce work through those officials most receptive to its ideas. In the end, it also meant that at the time of withdrawing from the collaboration the RF did not necessarily have the satisfaction of seeing the scheme attain the purpose for which it was created. Nor was there any hope that it ever would.

Concluding remarks

The RF in its eagerness to extend the benefits of scientific medicine and public health to as many communities as possible ardently pursued governments all over the world. The assumption underlying the RF approach was that it could be most effective and influential by getting governments to co-operate in the administration and implementation of its programs. Not all governments took the initiative to invite the RF or even evince interest in co-operating when approached by the RF. In such cases, the RF worked hard to receive formal invitations, which was but a mere formality. Invitations sometimes came after much persuasion. It is doubtful whether the RF officials were necessarily able to convey the fundamental principles underlying its various programs to the different authorities and agencies involved. Complains by RF officials about the lack of understanding about its programs amongst the political leadership, administration and the medical establishment were not entirely justified. This failure to grasp principles was the outcome of the manner in which the RF pushed its programs through with reluctant government agencies. It is true that the RF did "educate" the staff actually conducting the program in its practices and principles but experience showed they were not always very effective.

The assumption that since health and medical relief are matters of public good, philanthropic organizations and governments would have a commonality of interests and goals is fraught with many misconceptions. The attempt to reconcile differing concerns, interests and perceptions through negotiations may assure collaboration but not necessarily lead to close co-operation. This still leaves us with a few questions. Is the fact that in recent decades philanthropic organizations have been seeking partnership with community based and non-government organizations indicative of this failed relationship? Does its suggest loss of faith in governments' ability to deliver in a joint venture? The issue of how foundations and governments work out collaboration and a relationship is of continued and contemporary relevance. Is the relationship between foundations and non-government organizations any different? Do they face the same issues? Can philanthropic foundations truly expect to succeed where they failed with governments? Moreover, is the approach and attitude of the foundations towards non-government organizations different from that with regard to governments?

Notes

1. R. S. Morison, 7 September to 14 September 14 1952, Beruit, Lebanon, Office Diary, Record Group 12.1, Box 47, Rockefeller Archives Center (RAC).

2. Much has been written about the origins of the RF, its various constituent agencies and of its public health and medical programs and hence that part of its history need not detain us here. Here the reference to the RF is to the public health work of its major constituent agency, set up as the International Health Commission, later renamed the International Health Board and finally the International Health Division. For a very recent account of its international work see John Farley, *To Cast Out Disease. A History of the International Health Division of the Rockefeller Foundation (1913-1951)* (New York: Oxford: University Press, 2004).

3. For instance, the Government of Bombay having shown a great deal of reluctance through the 1920s and 1930s agreed to set up a demonstration health unit only in the early 1940s. See Shirish N. Kavadi, *The Rockefeller Foundation and Public Health in Colonial India. A Narrative History* (The Foundation for Research In Community Health: Mumbai/Pune, 1999).

4. Radhika Ramasubban, *Public Health and Medical Research in India. Their Origins Under the Impact of British Colonial Policy* (Swedish Agency for Research Cooperation: Sarec Report, 1982).

5. David Arnold, "The Rise of Western Medicine in India" *The Lancet* 348, (October 19, 1996) ; also see David Arnold, *Colonizing the Body, State Medicine and Epidemic Disease in Nineteenth Century India* (Delhi : Oxford University Press, 1993); for another account of public health in colonial India see Mark Harrison, *Public Health in British India. Anglo-Indian Preventive Medicine 1859-1914* (New Delhi: Foundation Books, New Delhi, 1994).

6. *Ibid.*

7 Christopher Baker, "Figures and Facts: Madras Government Statistics 1880-1940," in C. J. Baker and D. A. Washbrook eds. *South India: Political Institutions and Political Change 1880-1940*, (Delhi: MacMillan, 1975).

8. *Ibid.*

9. Roger Jeffery, *The Politics of Health in India* (Berkeley: University of California Press, 1988); also see for an account of public health administration under local self government bodies, Hugh Tinker, *Foundations of Local Self Governments in India, Pakistan and Burma* (London: Praeger, 1968).

10. V.R. Muraleedharan, "Rural health care in Madras Presidency," *The Indian Economic and Social History Review*, 24 (1987):328; also see V.R Muraleedharan, "Professionalising Medical Practice in Colonial South-India" *Economic and Political Weekly*, 27 (January 25, 1992);. V.R. Muraleedharan and D. Veeragahavan, "Disease, Death, and Local Administration. Madras City in Early 1900s" *Radical Journal of Health* (New Series) 1 (January 1995).

11. Kavadi, *The Rockefeller Foundation and Public Health in Colonial India.*

12. Rockefeller Foundation, "Distribution and control of hookworm disease in India, International Health Board, Indian" *Journal of Medical Research*, 10 (1922), 295-342.

13. John Ettling, *The Germ Of Laziness, Rockefeller Philanthropy and Public Health in the New South* (Cambridge: Cambridge University Press, 1981), 23.

14. In all, seven demonstration health units were initiated in India, of which four were set up in the Provinces and one at Delhi, the national capital. Two units were established in the independent Princely states of Travancore and Mysore. The following

were the health units, Neyyatinkara in Travancore (1931-1937), Closepet in Mysore (1935-1940), Partabgarh in the United Province (1932-1937); Poonamallee in the Madras Presidency (1935-1940), Najafagarh in Delhi (1936-1942), Sirur in Bombay Province (1938-1942) and Singur in Bengal (1939-1945). The last came to be attached to the All India Institute of Hygiene at Calcutta. Attempts at setting up health units at Golghat in Assam and at Pattoki in the Punjab were unsuccessful, see Kavadi, *The Rockefeller Foundation and Public Health in Colonial India.*

15. As quoted in Kavadi, *The Rockefeller Foundation and Public Health in Colonial India.*

16. W. C. Sweet "A Note on the Poonamallee Health Unit," RG 1.1 S464, Box 12, F96.

17. For an account of Madras politics of the time see, Christopher John Baker, *The-Politics of South India 1920-1937* (Cambridge: Cambridge Universlity Press, 1976).

18. *Ibid.*

19. W. C. Sweet "A Note on the Poonamallee Health Unit," RG 1.1 S464, Box 12, F96.

20. *Ibid.*

21. *Ibid.*

22. *Ibid.*

23. *Ibid.*

24. *Ibid.*

25. Jacocks to Ganapathy, 10 July 1935, RG 1.1 S464, Box 12, F96.

26. G.O. Ms. No. 2007/P.H. dated 1 Aug 1935, RG 1.1, S464 Box 12 F96.

27. Health unit Poonamallee Report for quarter ending, 31 Dec 1935, RG 5, S3 Box 206, F-India-Madras, Public Health Work.

28. Sweet, A Note, RG 1.1 S464, Box 12, F96.

29. Jacocks to Sweet, 28 May 1936 RG 1.1, S464, Box 12, F96.

30. Sawyer to Jacocks, 25 August 1936 RG 1.1, S464, Box 12, F96.

31. Sweet to Jacocks, RG1.1, S464, Box 12 F96.

32. Sweet's Note, RG 1.1, S464, Box 12 F96.

33. Sawyer to Jacocks, 30 Sept 1936 RG1.1 S464, Box 12, F96.

34. Jacocks to Sawyer 6 Oct 1936 RG1.1, S464, Box 12, F96.

35. Sweet to Jacocks 11 Feb 1937 RG5 S3 Box 202 F India- Annual Report 1936.

36. India-Semi Annual Report 1940 RG5 S3 Box 203.

37. Jacocks to Sawyer, 7 Oct 1940 RG1.1, S464, Box 12, F96.

38. India Annual Report 1940, RG5 S3 Box 203.

39. Government to Balfour, 8 April 1946 RG 1.1, S464 Box 12, F96.

40. Balfour to Gorman, 18 April 1946 RG1.1, S464 Box 12, F96.

Part Two

The Contemporary Scene

Chapter Eight

Community Based Primary Health Care: Empowerment and Equity

Carl E. Taylor and Henry G. Taylor

The theme of this conference is to review what we have learned in the past half century about the successes of philanthropic foundations in promoting health around the world. We will comment first on the Rockefeller Foundation's role in pioneering the first demonstrations of primary health care in developing countries. These projects built the foundation for the health systems of many developing countries and Central Europe. The Foundation has not received the recognition it deserves for many decades of low key but sustained breakthroughs promoting primary care and equity in health services.

A particular leader among the early Rockefeller giants was John Grant who is widely recognized to be the Father of Primary Health Care.[1] With Chinese colleagues such as Jimmy Yen and C.C. Chen in the 1920s and 1930s, he demonstrated remarkable success in the 400,000 population of Ting Hsien (now Ding Xian) county south of Beijing. This field research project became the model for Mao's system of Barefoot Doctors. From the 1960s to 1980s, China had one of the most equitable and cost/efficient health care systems in any country and it provided coverage for almost a quarter of the world's population. The whole system collapsed abruptly in the early 1980s when health services were privatized and commune work points were terminated by the Deng Xiao Ping economic reforms.

Following Ting Hsien there were second and third generation Rockefeller projects such as in Kerala and Sri Lanka led by Jacocks and in the Dutch East Indies, now Indonesia, led by Hydrick.[2] Around the world, many health systems based on demonstration primary health centers were established, including in Eastern Europe where they became the foundation for evolving Communist national health systems. John Grant became co-secretary with KCKE Raja in writing the Bore Committee Report in the late 1940s. This report laid the foundations for health systems on the Indian subcontinent. Another model was in Durban, South Africa where the very successful Pholella project was swallowed up by the Apartheid government. Kark and other leaders then went to Israel or the U.S. to start what became the Community Oriented Primary Care model under Lyndon Johnson's Great Society.[3]

Many years later the Rockefeller Foundation published monographs summarizing these experiences in a series of Bellagio Conferences on themes such as *Good Health at Low Cost.*[4] There was much speculation about why a state such as Kerala had achieved large scale and dramatic health improvements while still the poorest state in India.

It seems unusual that no one has studied in more detail the dynamics that caused these dramatic successes. They were sparked by early Rockefeller demonstration projects coupled with their emphasis on training health professionals. Analysis of this history might contribute to practical understanding of how primary health care can be implemented today. It has become evident that such infrastructure is essential to make all current health progress sustainable.

With initial Rockefeller funding in the 1950s, we developed the Johns Hopkins Narangwal Project in the Indian Punjab. In the 1960s and early 1970s, we demonstrated the impact of oral rehydration for diarrhea, in parallel with the pioneering field research at Matlab in Bangladesh. We also published the first demonstration of control of childhood pneumonia diagnosed by rapid and difficult breathing.[5] We also gathered definitive evidence on the synergistic interaction between malnutrition and infections, and on the importance of integrating family planning with maternal and child health.[6] Synergistic combinations of the interventions resulted in a rapid decline in child mortality by almost half.[7]

Jim Grant, John's son, led UNICEF during the Child Survival Revolution of the 1980s, which promoted national programs using new technical tools through mass, vertical programs, especially immunization and oral rehydration. I used to remind Jim that his father had said that one of the first principles of Primary Health Care was to integrate services, and he would say, "Yes, we will do that after we have implemented these simple programs."

Community Empowerment

We will now present some case studies described in more detail in our book, *Just and Lasting Change: When Communities Own Their Futures.*[8] The first principle in our book is the need for flexible balancing of control in a true partnership between bottom-up communities and top-down officials. A third partner is the outside-in expert who knows how to promote self-reliant community empowerment. There are now thousands of projects in all countries and under every economic system, which demonstrate that partnerships work well at the local level. They illustrate the truth of the phrase Peter Goldmark used in his presentation "creating a path by walking on it."

Development experts, however, have tended to reject these local experiences because they claim that rapid scaling up or extension did not occur. That is no longer true. We are learning how to Go To Scale quickly and with a systematic process. There are very few universal solutions, silver bullets, for the type of blueprint implementation traditionally attempted to extend successful pilot projects. The blueprint model fails to recognize that every community is different, with its own history, traditions, and problems.

We have described a universal process underlying equitable and sustainable community change, called SEED-SCALE. It allows each community to find its own appropriate solutions. Our case studies show that when local community empowerment takes off, the health services have trouble keeping up. It tends to occur where least expected. In any such scaling up, key people generate hope as they work for change, but each time the dynamics are different. Quantifying where and when the process happens is challenging, but the old proverb is true, "you know it when you see it."

We have observed that successful community projects often fail when officials and especially health service professionals consistently work in ways that produce dependency. That destroys the fragile self-reliance necessary to start effective empowerment. Officials and doctors have great difficulty learning how to support capacity building. They often become the main obstacles to empowerment when they could be the primary sources of support and encouragement. They start with precisely the wrong values and attitudes. Doctors take responsibility—but do not share it with patients; they assume they must be in control—not listening as a personal counselor. They tell people and assume that patients and families will do what they are told. Rigid, vertical, top-down management typically responds to crises or financial pressures by exerting the power of specialized knowledge and technology. However, these hospital-driven patterns of behavior, which might be appropriate in emergency care, clearly do not fit most health conditions. As a result, professional self-protection of specialty roles, privilege, and financial reward continues to permeate health care systems. This is despite abundant research demonstrating that the greatest impact comes from self-care in the home and behavior change in families.

Our case studies show an anachronistic generalization—that empowerment tends to emerge where health services are the weakest, with very few legal or bureaucratic constraints. In both developed and developing countries, ultimate success depends on prevention through behavior change. We believe that empowered people take responsibility for their own care in families and communities when participatory partnerships are encouraged and people share in ownership of their own care.

Case Studies

Tibet—Qomolangma (Mt. Everest) National Nature Preserve

About 15 years ago, the NGO Future Generations was invited by China to help develop the Qomolangma National Nature Preserve (QNNP) to protect the Tibetan side of Mt. Everest. It was one of the first major parks to have no wardens with guns. The 80,000 people living in the park, which is the size of Massachusetts, are the wardens. They protect the environment and we help with other development activities according to their priorities. The people quickly did their part in protecting forests and animals. They were so successful in environmental conservation that soon people were complaining that endangered Tibetan Wild Ass were so numerous that they devastated the barley fields and the snow leopards were carrying off too many sheep.

One of their highest unmet priorities was health because China's health care system never reached these remote villages. We helped with a training program for selected people from the villages, which they called Pendebas (Tibetan for a person who helps neighbors). We organized a month's training every summer and after four years had eighty-seven Pendebas working in their villages. They learned simple home remedies such as how to care for the number one cause of death, children's diarrhea, using cereal based oral rehydration made with supplies available in every home. As soon as symptoms started, mothers mixed the ORT solution—a beer bottle of boiled water, two matchboxes of Tsampa (roasted barley flour which is the main staple of Tibetan diets), and half a beer cap of salt. This was much more acceptable than trying to provide the WHO—UNICEF packet of ORS salts. The home-made mixture was readily available, cost nothing, and its more complex chemical balance actually served to reduce the duration of diarrhea, a fact mothers love since there is less cleaning up.

Four years after we started, the governors of the four QNNP counties moved into our training workshop and took over. They said we had created a serious political problem, which at first sounded as though they were going to terminate the program. However, the problem was that over 300 villages were demanding a Pendeba and they had calculated that it would take ten years for us to meet that demand at the rate we were going. Together we worked out a plan where the first groups we had taught could become trainers. Two years later, they had trained 320 Pendebas. Rapidly the responsibilities were expanded to include all social development activities and income generation, all done by volunteers.

Since then the empowerment has continued to expand rapidly under local Tibetan control. As the QNNP villages were covered, Tibetan officials in Lhasa insisted the program be extended to the new Four Great Rivers of Asia Nature Preserve that includes the headwaters of the Yangtze, Mekong, Salween, and Brahmaputra rivers. It is being extended once again to the Changtang or northern plateau. Forty-two percent of the land area of Tibet will be under conservation combined with social development. This has been achieved through a process of sustainable self-reliance and participatory partnership control.

Indian Himalayas—Arunachal Pradesh and Uttaranchal

At the eastern end of the Indian Himalayas, in what was NEFA (Northeast Frontier Area) under the British Raj, we have started a rapid process of Scaling Up. It is one of the most tribal areas of the world, with more than 30 distinct tribes. We started programs in three of the main tribal groups. Sille, where the Brahmaputra opens out to the plains, is the home of the Adis. Ziro, in a beautiful central valley system, is the home of the Apatanis. Both are the best-educated and economically advanced tribes. The third group, in a large area in the extremely rugged central mountains, is the Nishis. They were traditionally warriors, who tended to reject development activities and oppression of women was particularly severe. Local officials told us that they would be the last to develop, but we could try.

The reverse has been true. The Nishi women accepted innovation with immediate and rapid empowerment. It was a blessing in disguise that we started by talking about cholera since it showed the women how much they already knew. Before the first workshop, the Arunachal Director of Public Health had told us that Palin, the main town, experienced a cholera outbreak just two months before, and we should be careful about food and water. An epidemiological survey had found about thirty cases and eight deaths.

We first asked the women if the reports of cholera were true. The more than forty women in the workshop agreed with some passion and reported great fear. We said they could help us find out what really happened and then we would tell them how to control such problems themselves since the capital, Itanagar, and the public health services were eight to ten hours away by jeep.

They divided into eight neighborhood groups and discussed how the epidemic spread. Each group agreed on the date the first case in their neighborhood occurred, who became ill, their family, age, sex, etc. Then they traced each subsequent case with specific dates and data. Their memories were vivid because of concern for their own families. In half a day, they not only produced a precisely totaled and classic epidemic curve of 345 cases, but they focused especially on the thirty-five deaths. Reportedly illiterate women had tabulated age and sex, and patterns of cholera spread and death.

After congratulating them for knowing more about the epidemic than provincial epidemiologists know we told them that they could have prevented almost all the deaths. They did not have to rely mainly on intravenous rehydration at the hospital or on packets of Oral Rehydration Salts (ORS) from the health center. They learned they could prevent these deaths by using cereal and salt available in their homes, especially if they started as soon as diarrhea or vomiting appeared. In half a day they all became experts in making cereal based ORT for home use and ended up enjoying a relay race where individuals mixed up solutions and ran to give rehydration to teammates across the room.

The next day we took the rapidly growing group on a walk through Palin, the largest village in the valley, to learn how to prevent new cholera epidemics. In an apparent contradiction, they observed that the cholera cases were concentrated in Palin, as the administrative center of the region. It had been blessed several years earlier by the PWD (public works department) putting in a piped water system. The population was so remote that pipes had been simply laid on top of the very rocky ground and, as new roads and trails evolved, traffic went right over the pipes. These were now leaking generously with the water directly exposed to raw sewage in the gutters. Families had developed a remarkable spider web of repairs when leaking pipes no longer carried water.

Empowerment evolved rapidly as women went directly to the health department and the PWD to enlist their help in correcting the problem. The officials were not used to community women coming in groups with the self-assurance provided by detailed data and maps. This was the beginning of several greatly energized women's associations in the Nishi area who then began to prioritize multiple problems to improve living conditions. The women in greatest need were the ones leading the empowerment process.

Peru—CLAS (Comites de Administracion Salud)

About nine years ago, the Peruvian Minister of Health called to ask for help. The Shining Path (Sendero Luminoso) guerilla movement had moved out of the Andes into the slums of Lima. Mountain villages were saying to the government that they did not want health centers to be re-opened. One of us spent a month with a team from the Ministry talking with village groups. The groups said that the doctors from Lima did not like being assigned to remote villages and that the people could do a better job of running the health centers themselves. They said it was not the Peruvian army or American helicopters that had driven the guerillas out of the mountains. Rather, it was Ronderos or self-help groups, comprised mainly of women, who did not like having their teen-age boys kidnapped to become guerillas.

Our report suggested the government test the local idea of appointing committees to be responsible for their own health, with the health center doctor as chair. Fujimori, who was running for re-election as President at that time, issued a decree saying such committees could be formed. They made their own simple workplans based on a household health survey they did themselves. A year later over 150 committees were functioning and we spent another month with a ministry team evaluating the CLAS centers. Multiple follow-up evaluations then showed remarkable improvements in health outcome indicators, especially in 'soft indicators' such as equity. Again, it was mostly women's groups who were making sure services reached children and women in need.

Middle level health administrators then tried to stop the spread of the process by imposing regulations; they claimed these were necessary to protect the quality of care. The empowered communities then had to work out new participatory relationships, and there was much mutual learning on both sides. Now over 35% of the health centers in Peru have been taken over in a process called Shared Administration. A main source of hope is to follow such systematic scaling up to develop Modelos or Centers for Action Learning and Experimentation in each region. These can then design and test their own adaptations to local conditions and resources.

Summary of Scaling Up Process

Our book defines three principles of Scaling up:

1) *Three-Way Partnerships* between: *Empowered communities* in a system that ensures community ownership; *Officials* with legal support and regulations that guide local action; and *Experts* who bring the top-down and bottom-up together gently by providing experimentation for local adaptation and action learning.

2) *Evidence-based adaptation* building on local *successes* using participatory data gathering, local decision-making, functional analysis, role reallocation, self-reliant action, and evaluation.
3) *Community Empowerment* that depends on individual and family behavior change and community acceptance of new social norms using *annual workplans* to maintain momentum for new priorities and interventions.

Six criteria are used to measure progress in going to SCALE:
1) Equity

Equity is both a moral imperative and a practical way of making health services more effective. Good epidemiology starts with finding who in a population has specific health problems or is at risk of getting them. Only then does efficient control become possible. Surveillance for equity is a management tool that focuses health care on population groups where illnesses are concentrated.[9] This is almost always in the poor and disadvantaged families who tend to have multiple illnesses. Control measures can then be more efficient, compared to categorical programs which focus surveillance on only one problem. Nevertheless, communities by themselves seldom promote equity because local elites with power, property, and privilege benefit from ancient patterns of discrimination. Correcting disparities requires top-down and outside-in pressures which set standards that communities must meet.

2) Collaboration

The process that produces partnerships between the main action groups is equally needed for internal collaboration within communities. All communities tend to draw boundaries. Our definition of community is any group that shares something in common and has the capacity for joint action. However, distinctions between special interests and family groups still dominate communities and this leads to the chronic problem of factions that create internal stresses and competition. Empowerment brings factions together around a common vision and joint action, which is an important part of promoting conflict resolution and peace. We have found this is done best by women as they work together to improve the lives of children.

3) Sustainability

A health program is sustainable when it uses resources in ways that assure continued availability and it has three components in traditional communities. Cultural sustainability is when traditional values and social structures are protected. It is not just holding on to old-fashioned ways but preserving what is beautiful while adapting to new and better ways of living. Environmental sustainability requires measuring the preservation of natural resources and reducing pollution while balancing environmental protection with activities to improve the local economy. Economic sustainability faces three challenges: first, not passing costs to future generations; second, bringing in outside money usually produces dependency rather than self-reliance in communities; and third, is that partnerships should not measure development in terms of the money spent but focus instead on progress achieved towards community priorities.

4) Interdependence

The opposite of development is dependence, but the opposite of dependence is inter-dependence. Inter-dependence leads to webs of interaction opening access to people, resources, economic opportunities, technology, skills, and knowledge. As part of globalization, there is a tendency for communities to give up control, thinking they are gaining security as part of ever-larger organizations and market forces. Inter-dependence is an alternative approach that is also distinctly different from becoming increasingly isolated and losing the benefits of development. Modern communication channels provide opportunity for interactions that used to be impossible and communities need to gain the sophistication that is now monopolized by those in economic and political control. In health care, there is need for the people to make a fundamental shift from "compliance" to capacity building in habits and constant "adherence" to best practices for prevention.

5) Holism

In spite of much talk about integrated services, systems approaches and ecosystem perspectives, most governments maintain a sectoral focus in their community based programs. A current tendency is to narrow the focus even more from general health issues to specific health problems such as HIV/AIDS. The reason appears to be that outsider experts and donors can then more readily impose their priorities and control of money on communities, even though they have little awareness or interest in how local priorities can be fit together. We maintain that a genuine three-way partnership breaks through sectoral barriers and preconceptions to new horizons of sustainable problem solving.

6) Iteration

Seldom does any community-based action succeed on first trial. Instead, officials and experts need to learn how to "embrace error", and not just apply punitive controls. When something does not happen according to workplans, the person who identifies the problem, perhaps by making a mistake, should be congratulated and rewarded, because then solutions can be tried to make implementation better. Learning from mistakes and correction through repeated trials is how people naturally learn, and it is what social groups do best. Iteration helps communities to find their own way through local experimentation and adaptation, to solutions that were inconceivable a few years previously.

Seven Steps in Community Implementation, SEED (Self-Evaluation for Effective Decisions)

1) Building Capacity

Step One—Create a coordinating committee to mobilize the community and communicate with outside agencies.

Step Two—Identify the community's own past successes and strengths to build self-confidence and awareness of potential for change.

Step Three—Study other community successes and what has worked, if possible by visiting and talking with people who have produced positive change.

2) Choosing a vision

Step Four—Evaluate local situation realistically and instead of only getting opinions deliberately collect data on both problems and resources, especially to identify the community members with specific skills and motivation.

Step Five—Community dialogue about causes, gaps, and potentials for change and priorities. Agree on a workplan, functions to be performed and allocate assignments and roles to get care as close to homes as possible using community volunteers.

3) Taking Action

Step Six—Involve as many community members as possible from all factions. Respond to felt needs and build momentum.

Step Seven – Monitor activities and look for gaps and opportunities. Make midcourse corrections and test alternative options. It is less important to get everything right at the beginning than to learn, adapt, and keep improving.

Three dimensions of SCALING UP by the three partners

1) SCALE One— "Successful Communities Adapt, Learn and Empower". Community level capacity building demonstrates that a better life is possible and builds on successes to generate the momentum for community empowerment. This is the bottom-up component of the partnership where all communities develop ownership of the process.

2) Clusters of SCALE One communities make a commitment to help all the communities in their region in two ways. With participatory research, they identify priority problems, study causes, test options by practical problem solving, identify local resources both human and material, and demonstrate models of self-reliant success. Second is Learning by Doing where people who are successfully providing services guide others in the community and visitors from outside in learning how they can be successful. The purpose is to create self-confidence so others go out and try. Partners who are outside-in experts and who then develop local experts are the best to lead these activities. The goal is to develop a network of Learning Centers to pioneer development in each region.

3) SCALE Cubed—"Systems for Collaboration, Adaptive Learning and Extension"

Exponential or biologic models of scaling up – as distinguished from Blueprint or Explosion models—depend on having multiple growth nodes each of which actively promotes rapid expansion. This requires systems to help people change behavior and social norms, get new ideals and skills, and seek administrative and financial support. The first set of systems that need change is in partnerships and collaboration. Official and experts must learn new values to respect the people, build capacity, change policies, and regulations to permit bottom-up growth and share control with community leaders. A second set of systems is in adaptive learning, to encourage self-reliance rather than dependency and the ability to share knowledge and skills. A third set of systems is control of financing and administrative routines where truly transparent involvement of the people generates innovation and new patterns of support. For the basic principle that equity in care is essential for total population health, the fascinating finding is that this depends especially on having top-down imposed standards with which communities must comply. However, the system of surveillance for equity[10] is a straightforward management tool that can be used by the three-way partnership to achieve maximum impact.

Conclusion

Discontent with current practices where health services impose rigid programs that alienate peoples' participation are now producing cumulative demand for change. People are not things and previous Blueprint models of social engineering just do not work when people insist on controlling their own futures. Administrators who want to maintain control and keep their privileges and position have brushed off thousands of demonstrations of success in Community Based Primary Health Care. The have claimed that community based programs do not Go To Scale but this has been demonstrated in many successful programs to not be true. An anachronistic finding from our case studies is that Community Empowerment seems to emerge most successfully where health services are weakest and there are few doctors, showing the need for physicians to support rather than taking control. Communities must hold on to developing the capacity to own their health care and future development.

Notes

1. C. G. Seipp, *Health Care for the Community: Selected Papers of Dr. John B. Grant* (Baltimore: Johns Hopkins University Press, 1963).

2. J. L. Hydrick, *Intensive Rural Hygiene Work and Public Health Education of Netherlands India* (Batavia-Centrum: Java, 1937)

3. H. J. Geiger, "Community-Oriented Primary Care: A Path to Community Development," *American Journal of Public Health* 92 (2002): 1713-1716.

4. SB Halstead et al, *Good Health at Low Cost* (New York: Rockefeller Foundation, 1985).

5. C. McCord and A.A. Kielmann, "A Successful Programme for Medical Auxiliaries Treating Childhood Diarrhoea and Pneumonia," *Tropical Doctor* 8 (1978): 220-225.

6. Carl E. Taylor, et al, *Child and Maternal Health Services in Rural India—The Narangwal Experiment, Vol. 2 Integrated Family Planning and Health Care* (Baltimore: Johns Hopkins University Press, 1983).

7. A. A. Kielmann, et al, *Child and Maternal Health Services in Rural India—The Narangwal Experiment, Vol. 1 Integrated Nutrition and Health Care,* Baltimore: Johns Hopkins U. Press, 1983

8. D. Taylor-Ide and Carl E. Taylor, *Just and Lasting Change: When Communities Own Their Futures* (Baltimore: Johns Hopkins University Press, 2002).

9. Carl E. Taylor, "Surveillance for Equity in Primary Health Care: Policy Implications from International Experience," *International Journal of Epidemiology* 21 (1992): 1043-1049.

10. Carl E. Taylor, "Surveillance for Equity in Primary Health Care: Policy Implications from International Experience," *International Journal of Epidemiology* 21 (1992): 1043-1049.

Chapter Nine

The Need for Diversity in Models of Medicine

James Sensenig

I do not come to these proceedings as an expert in the accountability or evaluation of foundations or non-governmental organizations (NGOs). However, I've been asked to make some remarks on these subjects because of my background in naturopathic medicine and education, its clinical practice, the teaching of its history and philosophy, its political agenda for licensing and recognition, and because of my belief that naturopathic medicine represents an "outside the box" approach to healthcare and the challenges faced by clinical medicine both globally and within the United States.

I don't know to whom philanthropic foundations and NGOs are accountable in their promulgation of medical or health services. Presumably it would be to the constituencies that they serve, but it might very well be their own executive staff, boards of directors, or the philanthropists or other individuals funding the organizations. Accountability however is inextricably related to evaluation. When one knows to whom one is accountable, one must seek out the tools for evaluation which measure appropriate outcomes.

However, the expected outcomes themselves are defined by the philosophical model or paradigm through which, in this case, medicine is understood. The paradigm that has been dominant for most of the past century— "scientific" or biomedical medicine—is based on mechanistic philosophical assumptions, whereas the basic assumptions of naturopathic medicine and nearly all other systems of medicine including the indigenous medicine of many cultures are those of the vitalist school.

The principles of vitalism as articulated by Hippocrates can be summarized in three points. Firstly, health or a healthy state is defined as a balance among all systems and aspects of one's being. The corollary of which is that the adaptations which occur as this state of balance is disturbed are known as "disease." Secondly, the body and other living systems tend to move in the direction of balance and therefore the correction of the disturbance. The tendency or "force" which is invoked to explain this phenomenon is referred to as the "physis" or vital force, hence "vitalism." And thirdly, the process of change in and the symptoms expressed as a consequence of this change are different for each individual. From this it follows that the physician grounded in the vitalist model is interested in the reversal of the "disease" process by the restoration of health and not merely the treatment of symptoms. This may sound trite to some, but adopting a new paradigm—truly a new world view and not merely thinking more broadly—allows for clinical outcomes not known or expected by "scientific" medicine.

Three cases from my practice illustrate this point. One involves a four year old boy diagnosed with autism. According to "scientific" medicine this is a genetic disease for which there is no cure. Therefore the recommended treatment for this boy was medication and a behavioral modification program to manage the symptoms. When treated in the vitalist model extensive biochemical testing led to the correction of malabsorption secondary to subclinical inflammatory bowel symptoms, and to treatment for profound nutritional deficiencies and heavy metal toxicity. These imbalances were corrected over a three year period. As a result of which, the child is now in a mainstream educational setting and is functioning in an age appropriate manner.

The second case is that of a thirty year old female patient who complained of tension type headaches all day every day. Her headaches were worse at night, preventing her from sleeping restfully. She also complained of severe constipation, with a bowel movement once every two weeks or longer. She had been treated for years with "scientific" medicine but all of these treatments failed to alleviate the symptoms. In the vitalist model, her constipation was corrected through dietary changes, an exercise program, digestive enzyme supplementation, and restoration of normal bowel flora. A thorough history and examination also uncovered cervical spine dysfunction which was corrected by physical manual therapy. She improved incrementally over time and after several months was completely free of headaches.

In a third case, a thirty five year old woman was paralyzed after surviving severe head trauma from a fall on icy stairs. After the accident she was able to get around only in a motorized wheel chair and then only with help from others. Over a five year period with regular visits to a pain management center and with regular physical therapy she was able to walk with crutches, still in a great deal of pain. She was unable to care for herself and was considered completely disabled. After a series of physical medicine treatments including "cranial manipulation" and homeopathic remedies she was able to walk unaided and without pain. She is likely to make a full recovery and again be a productive member of society.

These examples are admittedly anecdotal and are included here to make a point. As Kuhn suggests, it is not possible to communicate across different paradigms.[1] Said differently, those whose world view is based on a particular set of assumptions, whether consciously articulated or not, may not be able to accept the premises of those who operate in a reality based on other assumptions.

Obviously, one who understands both paradigms can function in both. However someone who understands only one world view and does not "believe" in a different one cannot see what the other sees. Thus the person in the "old" paradigm has no conceptual framework or language to understand the "new" paradigm. What is possible and therefore acceptable within the context of any paradigm, is limited by its world view. Thus, one's world view with regard to medicine is no small matter. It defines not only what is expected of health care, but literally what is seen as possible.

With respect to evaluation of outcomes it is obvious that the foundations and NGOs involved in medical issues—whether healthcare services, education, or research—tend to use the model, and accept the assumptions of the currently dominant school of medicine. This limitation undoubtedly has a profound effect on philanthropy and health care.

This is not unlike the uncritical exporting of a particular model of government, whether or not it is desired or accepted by those who are presumed to benefit from it or with no thought of a larger examination of what else might be possible or more appropriate to the people involved, their conditions or their traditions. The point is that it is entirely possible, even to be expected, that other approaches will offer other solutions to problems facing health care on a global level. Competing medical systems based on different understandings of "health" and "disease" might well provide ideas that are unknown in the "scientific" model.

Without including the possible contributions of other systems of medical theory the evaluation of the conduct or mission of the foundations or NGOs becomes a closed circle. Evaluation is based on how well we might deliver certain services, impact illness, or reduce morbidity with respect to what the dominant model suggests is possible. This medical chauvinism may not only be self-congratulatory but also self-defeating if the mission of the foundation or NGO is truly to have the widest possible impact on the health of those receiving its services.

It is worth recalling the historical roots of this problem. At the beginning of the twentieth century two of the most influential foundations in the United States, those associated with Carnegie and Rockefeller, supported the development of the newly emerging "scientific" medicine over the vitalist schools that had developed within the rich tapestry of medicine in the United States. This had the political effect of selecting one system among many for propagation at the expense of the diversity of medical pluralism. As a result we now have a single system of medicine and related understanding of health, which for close to a century has received virtually all of governmental and philanthropic funds devoted medicine. Even as this system continues to dominate the landscape, its shortcomings are of increasing concern to most everyone in our culture. Having culled the diverse array of competitors from the landscape of medicine we now find ourselves with an approach to medicine which appears to be unable to serve our national or global needs.

Those of us involved in the practice, academic training, and philosophical discussions of other systems of clinical medicine would assert that they have something of value to offer our culture and those of the rest of the world. It seems axiomatic that encouraging and supporting the growth of these competing philosophical schools would strengthen our efforts—global and national—to improve health and medical care through diversity and competition. This might very well result in improved outcomes, lower overall costs, and increased satisfaction among consumers.

Many surveys suggest that as many as forty percent or more of American adults are seeking primary care from providers other than medical doctors and that they are willing to pay out-of-pocket for those services. The British Medical Journal (BMJ) reported that, "U.S. expenditure on complementary medicine rose in 1990-7 from $13 billion to $38 billion a year and twice as many consultations were with complementary practitioners as with mainstream family doctors."[2] Similar trends were reported for Australia and Great Britain. The same article reports that, "Most patients turn to complementary medicine out of frustration." A related article in the BMJ states, "It is now recognized that about half the population of industrialized countries regularly use complementary medicine."[3] Further, if one assumes that most people in the non-industrialized countries do not use or do not have access to "mainstream family doctors" it is easy to conclude that "scientific" medicine is not meeting the needs of the majority of the world's population.

The future of medical research will focus on examining the outcomes of whole systems or whole disciplines as opposed to examining individual parts or therapies taken from a discipline. This is because the synthetical or vitalistic systems of medicine not only understand symptoms to be manifestations of underlying imbalances, but also tend to use therapeutic interventions which correct those imbalances rather than simply suppress symptoms. In these models several persons with similar symptoms or the same disease may require several different interventions. Said differently, the emphasis will be on the patient who has the disease rather than on the disease that has the patient. This is compatible with the emerging discussion of applying complexity science to medical thinking as opposed to the simplistic mechanistic model now in common use.

Evaluating vitalistic therapies in the context of their individual medical models is not the same as incorporating various treatment modalities of non-conventional systems into "scientific" medicine. Doing the latter may slightly expand the therapeutic interventions of the dominant school, but it does not enrich the landscape of diversity. Such co-optation narrows rather than expands the choices, may decrease therapeutic efficacy, and may be dangerous. Again from the BMJ, it is reported that integrating traditional Chinese medicine into "modern" medicine resulted in a "lesser system." An Australian study found that acupuncture practiced by medical doctors resulted in more than twice as many "adverse events" than that of practitioners trained in traditional Chinese medicine.[4]

We need medical pluralism for the same reasons that we need biodiversity to maintain the genetic pool. I would suggest that because there is now a surplus of medical doctors in the United States and since "scientific" medicine enjoys unequalled favoritism at the hands of our government, that those who wish to make a difference in the world with private funds consider directing support to the promising and organized competing medical systems that have been restrained by funding, regulation, and politics. Those who do so, in my opinion, will be on the cutting edge of the revolution of healthcare.

Competition among medical systems would likely, over time, reduce morbidity and mortality, decrease healthcare costs thereby improving national economies, and point the way to sustainable medical models. True medical pluralism is certain to have far-reaching benefits for global healthcare.

Notes

1. Thomas Kuhn, *The Structure of Scientific Revolutions* 2nd ed., (Chicago: The University of Chicago Press, 1970).
2. L. Rees and A. Weil "Integrated medicine" *British Medical Journal* 322 (2001):119-20.
3. G. Bodeker, "Lessons on integration from the developing world's experience," *British Medical Journal* 322 (2001): 164-7.
4. A. Bensoussan and SP Myers, *Towards a Safer Choice: The Practice of Traditional Chinese Medicine in Australia* (Macarthur: University of Western Sydney, 1996).

Chapter Ten

Agenda-Setting Diseases

Ronald T. Rozett

What are agenda-setting diseases? This phrase, used by Professor Stephen Latham of the Quinnipiac University School of Law during discussions planning this conference, identifies diseases which set the political, economic and public health agendas for the countries in which these diseases are prevalent. For purposes of our discussion, agenda-setting diseases include AIDS, tuberculosis and malaria. They have caused social fragmentation, economic decline, and political polarization in countries where they are endemic and, as such, should be considered to be security as well as humanitarian issues deserving of robust research support.[1] Such diseases have major impacts on countries, regions, and cultures, potentially leading to the collapse of poor nations. At a conference of the G8 nations in July 2000, a communiqué released at the end of the conference stated that "infections and parasitic diseases, most notably HIV/AIDS, tuberculosis and malaria . . . threaten to reverse decades of development and to rob an entire generation of hope for a better future."[2]

Consequently, philanthropic foundations have pledged billions of dollars for global health initiatives, most involving infectious diseases. Whether this money has had, and will continue to have, on balance, a beneficial impact on the health of populations is the subject of this conference.

A brief review of the historical context of contagious disease indicates the power of epidemics to alter the social, economic, and political realities of life for vast numbers of the poor who are disproportionately affected by them. Bubonic plague, springing from its reservoir in wild rodents and spread by fleas, is one of the diseases referred to in the Bible. The first well described plague epidemic occurred in the 6th century AD in the Roman Empire where it is estimated that the epidemic killed half the population. In 1348, a plague epidemic, originating in eastern Asia, spread across Europe, ultimately reaching England. In that pandemic, bubonic plague killed an estimated twenty-five million people, including half the population of England.

Major agenda-setting contagious diseases include cholera, yellow fever, smallpox, typhoid fever, typhus, poliomyelitis, sleeping sickness (African trypanosomiasis), and influenza, but for purposes of this presentation I will focus on tuberculosis, AIDS and malaria.

Rene Dubos, Amartya Sen and Paul Farmer have eloquently discussed the social and economic determinants of disease. Disparities of risk and outcome are part of the complex interactions between biological and social realities. Disparities are particularly observable when new and effective therapies are unavailable to many of those in need. Social inequalities determine both the distribution of modern plagues and the clinical outcomes among the afflicted. Income differences in a society are among the most powerful determinants of health. To quote Farmer, "most of the bottom billion have no modern medical care at all. Most of the boundaries we create to shield ourselves from this reality are ones we create ourselves—we create them to lessen *our* pain, not the pain of others." Throughout history, it is clear that the great epidemics that have spread across large geographic areas, and indeed around the world, affect poor and marginal members of society disproportionately.[3]

Tuberculosis

Tuberculosis has been called the Great White Plague to distinguish it from bubonic plague, the Black Death. It is caused by a slowly growing bacterium, *Mycobacterium tuberculosis*, which can remain dormant within infected people for years or decades. Once infected, people do not necessarily become clinically ill. The infection is contained by the body's immune system, becoming active in the presence of factors which reduce immune resistance. In other words, the causative bacterium, in the words of Rene Dubos, is a necessary but not sufficient agent of the disease.[4]

In the U.S. in the year 1850 tuberculosis mortality rates exceeded 500 per 100,000 inhabitants. Even in the early 1900s, tuberculosis was the most common cause of death in the U.S. Starting in the late 19th century in this country and in other industrialized countries, tuberculosis began to decrease as a major cause of death, long before the appearance of effective medications to treat it. Death rates dramatically decreased tenfold in less than a century. Although this trend was interrupted by two world wars and by periods of social upheaval following the wars in Europe, the decline continued when order and economic development returned.[5]

The tuberculosis epidemics in the 19th century which prevailed throughout the industrialized world were caused at least in part by long hours working in difficult conditions, poor nutrition, marginal housing, the absence of adequate heating, and the generally poor standard of living among the working class during that period. As the standard of living for workers improved, tuberculosis mortality gradually decreased.[6]

Tuberculosis has always disproportionately affected poor people. In 1900 in the U.S., the death rate for whites was 200 per 100,000 persons; while it was

400 per 100,000 for blacks. And even among whites, the case fatality rate for those with clinical disease was seven times higher in unskilled laborers.[7]

WHO declared tuberculosis a global emergency in 1993 because of its high prevalence in Third World countries. Tuberculosis had become largely invisible to Americans because of its persistent patterns of differential susceptibility in the last few generations, largely sparing the well to do. But now, mostly because of the AIDS epidemic, tuberculosis has "reemerged," the term used by Paul Farmer, from the ranks of the poor. In New York City, the number of tuberculosis cases by the early 1990s had nearly tripled in 15 years, so that the case rate there of 222 per 100,000 exceeded that seen in many developing countries.[8]

Between 2000 and 2020, it is estimated one billion people will become infected around the world, more than 150 million people will become sick and contagious, and thirty-six million people will die of the disease[9]—staggering numbers for a disease almost forgotten in the U.S.

Once infected the poor are more likely to develop active disease because of the presence of other conditions that weaken the immune system, e.g., malnutrition, AIDS, other severe infections, fatigue, and alcohol and drug addiction. Left untreated each person with active tuberculosis infects approximately ten to fifteen people. Nearly one percent of the world's population is newly infected each year. Of these, nearly five to ten percent becomes contagious at some point in their lives.

Multi-drug resistant strains of *Mycobacterium tuberculosis*, MDR-TB, have appeared in virtually every country in the world as a result of frequently inconsistent compliance with tuberculosis treatment regimens. Individuals with this resistant form of tuberculosis then expose others who consequently suffer a primary infection with the resistant strain. Should they become sick, their infection becomes more difficult and more expensive to treat.

Social circumstances largely determine clinical outcomes in tuberculosis. There are large-scale forces that make monotherapy and erratic drug ingestion much more common in settings such as India, Haiti and Harlem than in affluent communities. Strong predictors of compliance with drug treatment which must be sustained over many months are fundamentally economic, not cognitive or cultural argues Paul Farmer. Those *least likely* to comply are those *least financially able* to comply with drug regimens.[10] This is consistent with the well-established association between disadvantage and disease.

In response to the increasing prevalence of MDR-TB, WHO initiated DOTS programs, Directly Observed Treatment, Short Course, around the world reaching fifty-five percent of the global population. Nearly $500 million was spent on DOTS programs in 2000 in the 22 so-called high burden countries which account for eighty percent of all new tuberculosis cases in the world.[11] DOTS is WHO's recommended tuberculosis control strategy using relatively available first-line tuberculosis medications: isoniazid, rifampin, pyrazinamide, and, in seriously ill patients, ethambutol. DOTS programs require a sustained commitment from governments to deliver effective laboratory-based diagnostic capability and free standard treatment under direct observation by trained personnel three times a week on alternate days. An intensive treatment phase lasts two months, followed by a continuation phase of four to six months. INH/rifampin or INH, rifampin and ethambutol are prepared in weekly packs for patient use with the first dose of the week observed. The program is dependent on a secure drug supply and systematic monitoring and evaluation.[12]

The problem with the DOTS approach is MDR-TB does not respond to these drugs. DOTS programs have produced an average global cure rate of 82% in 2000, but a rate of only 72% in Africa[13]. Treatment failures represent, at least in part, the ineffectiveness of standard medications in eradicating MDR-TB.

In India, 2 million persons develop tuberculosis every year. This constitutes 25% of the world's new tuberculosis cases. It causes 420,000 deaths annually, more than malaria and all other infectious, tropical diseases in the country and this has occurred despite India's best efforts for forty years. During this time the incidence and prevalence of tuberculosis has not changed substantially. In 1992, the Indian government established the Revised National Tuberculosis Control Programme using the DOTS multidrug regimen. This revised program was implemented widely in 1998 for 40% of the country's population, 440 million people. In India, 83% success rates were observed using the DOTS program within a few years of its inception in areas where it was implemented, but the success rate among previously treated patients was only 71%.[14] Despite the substantial presence of MDR-TB, as a result of the DOTS treatment program the tuberculosis death rate dropped from 29% to 4%.[15]

However, in India tuberculosis medications can be obtained without a prescription, leading to indiscriminate and inconsistent utilization of these drugs, contributing to the appearance of drug resistant strains. In addition, economic hardships and drought have led to large-scale migrations which interfere with the completion of a full course of drug therapy. Finally, India has not been able to implement a transportation support system to get patients to program sites.[16]

MDR-TB in India now accounts for 1-3% of new tuberculosis cases. In Russia the comparable number is 10-15% and in New York City it is 7%.[17] This poses a real threat to the sustained success of the DOTS program in India, especially since the second line tuberculosis medications, effective against MDR-TB, are not available at affordable prices in India. It is about 100 times more expensive to treat MDR-TB and, unfortunately, treatment is associated with more serious side effects.

With the appearance of AIDS in India, now constituting about 4 million cases, and its synergistic effect on the prevalence and severity of tuberculosis, efforts to control tuberculosis will be even more difficult. Today, about 50% of AIDS patients in India also have tuberculosis.

India's tuberculosis control program is critical to global tuberculosis control, including control of the disease in the U.S. Fifty percent of all new cases in the U.S. occur among foreign born persons, about ten percent of whom are from India.[18]

In summary:

1. An estimated two billion persons, 33% of the world's population, are infected with *Mycobacterium tuberculosis;*
2. Eight million people develop active tuberculosis every year;
3. Tuberculosis is the second leading cause of death among adults in the world;
4. Between two and three million people die of tuberculosis annually; and
5. Tuberculosis is the leading cause of death from infectious disease among the world's young people.[19]

Acquired Immunodeficiency Disease

The disease which eventually became known as AIDS was first reported in the U.S. in June 1981 in an outbreak of a rare form of pneumonia, *Pneumocystis carini*, or PCP, among five young gay men in Los Angeles.[20] Additional reports soon followed from New York City and San Francisco.[21] These outbreaks were quickly recognized as examples of cellular immune dysfunction associated with sexual contact. By early 1983, other risk factors were identified by the Centers for Disease Control, the CDC, all of which had in common exposure to the blood of previously infected people sharing needles among IV drug abusers, recipients of blood transfusions and blood products (especially common in hemophiliacs), maternal to infant transmission, and occupational exposures, primarily in health care workers. In March 1983 the CDC issued its first set of prevention recommendations.

AIDS became the most thoroughly and rapidly researched infection in human history. Within three years of its recognition as a new disease entity, the unique retrovirus causing AIDS was discovered; a sensitive and specific diagnostic test for the presence of antibodies produced by infected individuals was developed; and the test was used in the field to survey populations and to screen blood donors.[22] It became clear that HIV infected people could remain asymptomatic for up to ten years, acting as silent carriers of the disease. So the distinction between HIV infection and AIDS became an important one. The remarkably rapid development of antiretroviral drugs followed and by the mid-1990s combination therapy with potent antiretroviral medications was well established as very effective treatment for AIDS. This became known as highly active antiretroviral therapy or HAART. It is very expensive, costing between $10,000 and $15,000 a year per patient in the U.S. The use of HAART, although not curative, has led to a dramatic decrease in mortality among patients in industrialized countries, making AIDS a manageable chronic disease by effectively suppressing the virus infection.[23] This is unique in the history of fatal virus infections among humans.

In summary, AIDS in the U.S. and in many other industrialized countries has now stabilized in contrast to the situation in developing countries. One of the important factors leading to the control of the AIDS epidemic in the U.S. is that since 1981 total NIH funding for AIDS research and treatment programs has exceeded $20 billion.[24] By contrast, in 2001 the global HIV and AIDS epidemic had become the fourth leading cause of death in the world. It is a global pandemic disproportionately affecting disadvantaged and marginalized members of society: poor people, sex workers, and IV drug users.[25] Thirty six million people have been infected in the 20 years since AIDS was identified. It has caused an estimated 20 million deaths, with tuberculosis being the principal cause of death for AIDS patients worldwide.

AIDS in Africa

The region of the world most severely affected by HIV and AIDS today is sub-Saharan Africa. About 25 million people are infected with HIV, with four million new cases in 2000 alone. The average prevalence rate is 10% among people in 16 African countries in the age group fifteen to forty-nine; but in Botswana, 36% of the adult population is infected. Life expectancy has decreased by ten years and infant mortality rates have doubled throughout sub-Saharan Africa as a direct result of the AIDS epidemic.[26]

AIDS has become the leading cause of death in Africa, surpassing malaria in many countries. More than sixteen million people have died from AIDS in sub-Saharan Africa. Tuberculosis deaths in some African countries have increased fourfold to more than 400 per 100,000 as a direct result of HIV/AIDS. In some countries, 70% of persons with active tuberculosis are HIV positive.[27]

As staggering as this is, the full impact of AIDS in sub-Saharan Africa has yet to be felt. The disease's impact on families, social systems, and national growth and development is almost incalculable in some countries. Afflicting young people during their most productive years, AIDS has produced severe shortages of teachers, doctors and nurses. Large numbers of orphaned children and widows with no means of support have appeared in many countries. AIDS has shredded the social and economic fabric of large parts of sub-Saharan Africa, destroying all hope of improvement in the standard of living in this part of the world for the foreseeable future.

Despite great obstacles, in countries like Uganda intensive prevention programs have slowed AIDS transmission. In Uganda, AIDS prevalence dropped from fourteen percent to eight percent in the two years following the implementation of an aggressive education and prevention program which included strict regulation of commercial sex, intensive condom promotion, and the treatment of other sexually transmitted diseases which enhance the transmission of AIDS.[28]

However, in sub-Saharan Africa, AIDS is truly a disease setting the public agenda: nothing of consequence can be accomplished in these countries before the AIDS epidemic is somehow contained.

AIDS in other Geographic Areas

The incidence and prevalence of AIDS are also increasing in China and India. Given their sizable populations, the consequences of this disease for these countries could be as catastrophic as in Africa. Both countries were slow to respond to the emerging epidemic and both will face the same devastating consequences currently seen in Africa unless prevention and treatment interventions are intensified.

In Eastern Europe and Central Asia, IV drug abuse was responsible for the introduction of AIDS into countries like Ukraine which has 90% of the region's reported cases. Once the disease was introduced, it was propagated in the usual ways: sex, blood transfusions, and occupational exposures.[29]

There were about 1.4 million cases of AIDS in Latin American and Caribbean countries by the end of 2000. The prevalence rate of 2.1% in the Caribbean is second only to the rate of sub-Saharan Africa.[30] As is the case with every other epidemic infectious disease, differences in infection rates are accounted for by differences in economic and political stability, migration, poverty and unemployment, illiteracy, and drug use.

Success in stemming the AIDS epidemic even in countries with limited resources has been achieved through strong, high level government leadership in addressing the AIDS problem and with active community involvement. Successful programs emphasize the following measures:

- Testing the blood supply
- Promoting condom use
- Educating young people beginning with children in elementary school
- Increasing access to chemical dependency treatment and distributing clean needles to IV drug abusers
- Administering antiretroviral agents with a high priority on preventing mother to infant HIV transmission
- Treating STDs
- Testing for HIV with individual patient counseling—but the stigma of being HIV positive must be overcome to do this effectively[31]

As with MDR-TB, however, the cost of treatment for patients with AIDS is prohibitively high in developing countries. This has led to health policy strategies declaring HAART treatment "cost ineffective." In Paul Farmer's opinion, this represents the conventional establishment view which values human life differently based on measures of poverty and inequality. Inequality is an important co-factor in this pandemic.[32] Stopping AIDS will require a more ambitious political, economic and public health agenda than most countries are currently prepared to fund. To accomplish this, developing countries will have to significantly shift their spending from purchasing arms to implementing effective AIDS prevention and treatment programs.

In Farmer's view, social inequalities determine not only the dimensions of the AIDS pandemic but also the clinical outcomes once individuals become clinically ill.[33] The high cost of antiretrovirals has put them beyond the reach of most HIV infected people in the world. Farmer believes that efforts to make these medications available should be the highest priority of the world's traditional donors of health care funds, including foundations. Both the NIH and the CDC global AIDS programs will fall far short of their goals if access to antiretroviral medications is not addressed aggressively. Inevitably, this means difficult negotiations with the pharmaceutical industry, dealing with countries squandering scarce resources on arms, and, perhaps even more difficult, confronting the countries that sell arms to developing countries.

Governments in Africa and those of other developing countries face many challenges competing for their scarce resources. They simply cannot mount an adequate response to fight the AIDS pandemic without outside assistance; but at the moment, HIV is spreading three times faster than is the funding to control it.[34]

UNAIDS created an International Partnership Against AIDS in Africa providing support for indigenous expertise and institutions in implementing prevention programs. It is providing support for the development of the public health infrastructure needed to mount effective prevention efforts. But it is tacitly writing off a generation of infected people for whom these efforts will be too late with resulting devastating social and economic consequences .[35]

The AIDS epidemic now constitutes a global threat to the security and welfare of the U.S. and it is in our enlightened self-interest to vigorously assist other countries in controlling it. It can, and most likely will, lead to political instability and ethnic conflicts around the world;[36] while it undermines decades of efforts to improve the standard of living in Third World countries. It is, in brief, nothing short of a global catastrophe equivalent to a major global war.

To quote Paul Farmer once again describing the AIDS pandemic, "Their sickness is a result of structural violence: neither culture nor pure individual will is at fault. Rather historically given, and often economically driven, processes and forces conspire to constrain individual agency. Structural violence is visited upon all those whose social status denies them access to the fruits of scientific and social progress. . . . it produces a differential political economy of risk."[37]

Malaria

Malaria is the most prevalent vector-borne disease in the world, endemic in 92 countries containing more than 40% of the world's population. Annually, 300 to 500 million new malaria cases are reported, 90% of them occurring in Africa. Every year 1.5 to 2.7 million people die of malaria, 50% of them children younger than five years.[38]

Malaria is caused by four different types of a protozoan parasite, *Plasmodium*, which is transmitted through the bite of an infected female Anopheles mosquito. The involvement of an insect vector makes malaria control challenging in ways that TB and AIDS are not.

Historically, malaria has had a great impact on societies as a result of its insidious and debilitating effects on infected individuals, especially children. Relapsing fevers and the prolonged, chronic illness associated with malaria during infancy and childhood slow development and the ability to learn.

Malaria was endemic in large areas of the U.S. as recently as the 1930s. In 1914, about 600,000 cases occurred in the U.S. In 1998, only 1227 cases were reported, most of them in individuals from other countries.[39] Its disappearance in the U.S. and in other countries had little to do with personal medical care. It was largely a result of improved socioeconomic conditions, vector control efforts, including water management to reduce mosquito breeding sites, the use of window screens, and case management to interrupt malaria transmission by identifying and treating individuals with malaria.

Malaria is treatable and preventable with a variety of medications, but as with tuberculosis, treatment has produced strains of the parasite resistant to common medications. And, once again, medications effective against resistant strains of the parasite cannot be afforded by the people in endemic areas who need them the most.

Malaria and Population Movement

The world's urban population is growing at four times the rate of the rural population ("urban pull") contributing not only to the spread of malaria, but to the spread of tuberculosis and AIDS as well. Sub-Saharan Africa is the most rapidly urbanizing region in the world; while the urban population of India has doubled in the last two decades.[40]

Migratory laborers in agriculture and mining and international air travelers also contribute to the spread of malaria as well as other epidemic diseases such as cholera, typhoid fever and tuberculosis. Mass movements of populations caused by natural disasters, like droughts and floods; man-made disasters like civil or international conflicts; and pressure on scarce resources leading to migration of people from rural areas to urban areas also contribute to the spread of malaria. Malaria is spread when people migrate from endemic areas to areas where the disease has been eradicated.[41] It represents one of the most common causes of death among refugees.[42]

Man-made changes in the environment, such as deforestation and the development of irrigation projects, also contribute to conditions for mosquito breeding and increase the chances of malaria transmission. Workers in areas being developed for human agricultural and industrial use are at increased risk of contact with infected mosquitoes.[43] And finally, population movement has accelerated the spread of drug resistant strains around the world.

As with tuberculosis and AIDS, we need to devote more resources to controlling malaria. Perhaps, most important, is devoting more resources to the development of an effective malaria vaccine, efforts unsuccessful up to the present time because of the parasite's remarkable adaptive abilities. Given our growing understanding of the parasite's genome, this research could be fruitful in the foreseeable future.

In conclusion, three agenda-setting diseases, tuberculosis, AIDS and malaria, dominate the political, social, economic and public health realities of most developing countries. It is highly unlikely that sustained improvements in standards of living are attainable in most developing countries without controlling the spread of these endemic diseases and effectively treating individuals afflicted by them. At the same time, it is clear that developing countries do not have the resources to mount effective campaigns against these diseases without significant help from the wealthy countries of the industrialized world. Quite aside from the laudable humanitarian rationale for this, it is certainly also in the best interests of industrialized countries to assist developing countries struggling with AIDS, tuberculosis and malaria for two reasons: to reduce the danger of political and economic instability caused by these diseases in large parts of the world; and to protect their own populations against the health risks posed by these formidable diseases.

Increasing the availability of effective medications at affordable prices subsidized by industrialized countries is a beginning in tackling these complex public health issues. But much more is needed. Providing funding for education, for improvements in public health infrastructure and for sustainable economic development consistent with the indigenous cultures of recipient countries is critical to achieving permanent progress against these diseases. Countries must address the social and economic factors contributing to and perpetuating the transmission of AIDS, tuberculosis and malaria if they are to favorably change the environment that currently sustains these endemic diseases. Otherwise, medication alone will produce only transient improvement.

It is imperative that we view these agenda-setting diseases as threats to international security as important as the more obvious threats to international peace. In this context, redirecting a sizable amount of our international aid resources from military priorities to public health priorities is long overdue.

Notes

1. AC Fauci and GK Folkers, "The AIDS Research Model: Implications for Other Infectious Diseases of Global Importance," *Journal of the American Medical Association*, 286 (2001): 460.
2. *Ibid.*, 461.
3. Paul Farmer, *Infections and Inequalities: The Modern Plagues*, (Berkeley: University of California Press, 1999), 15-16.
4. R. Dubos, *Man Adapting*, (New Haven: Yale University Press, 1965), 164.
5. *Ibid.*, 169.
6. *Ibid.*
7. Farmer, *Infections and Inequalities*, 185.
8. *Ibid.*, 198.
9. World Health Organization. *Tuberculosis: Fact Sheet No. 104.* (Geneva: World Health Organization, 2002).
10. Farmer, *Infections and Inequalities*, 253.
11. World Health Organization, *Tuberculosis*.
12. Centers for Disease Control and Prevention, "Progress Toward Tuberculosis Control—India, 2001" *Morbidity and Mortality Weekly Report* 51.11 (2002):229.
13. World Health Organization, *Global Tuberculosis Control: Surveillance, Planning, Financing—WHO Report* (Geneva: World Health Organization, 2003).

14. Centers for Disease Control and Prevention, "Progress Toward Tuberculosis Control," 231.

15. *Ibid.*

16. *Ibid.*, 232.

17. *Ibid.*, 232.

18. World Health Organization. *Tuberculosis: Fact Sheet 104.*

19. *Ibid.*

20. Centers for Disease Control and Prevention, "*Pneumocystis* pneumonia, Los Angeles," *Morbidity and Mortality Weekly Report* 30(1981): 250-252.

21. Centers for Disease Control and Prevention, "Kaposi's Sarcoma and *Pneumocystis* Pneumonia Among Homosexual Men—New York City and California," 30 (1981): 3-5.

22 Fauci and Folkers, "The AIDS Research Model," 458.

23. *Ibid.*, 459.

24. *Ibid.*, 458.

25. Centers for Disease Control and Prevention, "HIV and AIDS, 1981-2000," *Morbidity and Mortality Weekly Report* 50 (2001):434.

26. *Ibid.*, 435.

27. *Ibid.*, 437.

28. *Ibid.*, 438.

29. Farmer, *Infections and Inequalities*, 52-56.

30. *Ibid.*, 52.

31. Centers for Disease Control and Prevention, 2001, 438.

32. Farmer, *Infections and Inequalities,* 57.

33. *Ibid.*, 52.

34. Fauci and Folkers, "The AIDS Research Model," 460.

35. Farmer, *Infections and Inequalities,*79.

36. Fauci and Folker, "The AIDS Research Model," 460.

37. Farmer, *Infections and Inequalities* 79.

38. P. Martens, and L. Hall. "Malaria on the Move: Human Population Movement and Malaria Transmission," *Emerging Infectious Diseases* 6 (2000): 103.

39. Centers for Disease Control and Prevention, "Malaria Surveillance—United States—1998," *Surveillance Summaries* 50 (2001): 1-18.

40. Martens and Hall, "Malaria on the Move," 104.

41. *Ibid.*, 103.

42. *Ibid.*, 106.

43. *Ibid.*, 105.

Chapter Eleven

Health Systems Reform—Seeking the Holy Grail

Thomas Rathwell

There has been considerable discourse in the last ten years or so arguing that the traditional bureaucratic approach to public sector government and management is not working and that a different paradigm is required, one based on the discipline of the market. In *Re-inventing Government*, Osborne and Gaebler (1992) argue that governments should not provide public services directly, rather they should leave provision to other agencies, such as the community, voluntary or private sectors who have a different ethos—one that is entrepreneurial and transformative. As Osborne and Gaebler aptly phrase it, the role of government is to "steer and not row." It is this concept that governments should 'steer and not row' that has been the driving force behind much of health sector reform over that last decade. Hood (1991) defines it as 'new public management.'

New public management has two distinct components: the adoption of production engineering techniques into public sector organizations, and application of market-based approaches to the delivery of public services. As Walsh (1995) notes: "this approach to management involves the creation of a core-periphery model of organization. . . . At the core are the central strategists, while at the periphery are those who deliver the services" (xiv). New public management hinges on the assumption that the economic incentives inherent in a market-based competitive environment would deliver the collective benefits of greater efficiency and higher quality services.

The discussion begins with an overview of public sector reforms and their specific manifestation in the health sector. The concepts of accountability and evaluation are then discussed as a cornerstone of the reform process. This is followed by an assessment of the impact of reform on hospitals as provider organizations.

Public sector reform

Public sector reforms can be grouped into three stages. The first stage consisted of the divestment by governments of such commercial enterprises as airlines. The second stage consisted of the privatization of public utilities such as electricity, telephones and water suppliers. The third and most extensive stage was the 'corporatization and/or marketization' of public social services. This latter phase has been most evident in the health sector where governments of all political persuasions have sought to transform the way in which health care is purchased and delivered. The reforms have been more pronounced in some countries than in others. One of the most significant results of the health care reform process which is well-documented in the literature has been the effect on purchaser organizations. The impact on provider organizations has been less well-documented. Most of the literature which comments on health reform from the provider point of view does so from a political perspective. Much of this is concerned with the fundamental transformation of provider organizations in the British NHS from directly managed public entities to quasi-independent providers (Flood 1997).

In the health sector, the market-based competitive environment is manifest primarily in the purchaser-provider split. The corporatization—marketization reform of health services generally has taken one of three different but related forms: internal market, managed competition or managed care. The *internal market* occurs where publicly funded providers are required to compete for service contracts from purchasing agencies. For example, the modality of reform in the United Kingdom was the internal market where the health system is funded by taxation and providers are state-owned entities. *Managed competition* refers to the system whereby insurers compete for customers on the basis of price and quality. In this system a form of internal market may operate as providers compete for service contracts from insurers. In Germany and the Netherlands, with funding mainly through social insurance mechanisms and a mix of state and private (not-for-profit) providers, the primary mode of reform was managed competition. *Managed care* is used to describe any arrangement whereby a purchaser, through the contractual process, seeks to affect the price, quality and quantity of services to be delivered. There are no pure forms of these models in operation. The extent to which any one of these reform mechanisms was implemented is a function of the manner in which the health care system is funded and the existing array of provider agencies.

It seems, however, that the reality does not support the theory (Lane 1997). Walsh (1995) too is sceptical about the value of uncritically applying business principles to public sector organizations. There are significant disadvantages with markets just as there are with bureaucracy. Both have something to contribute, principally because they are derived from very different, but not necessarily incompatible, value systems. Thus, Walsh (1995) concludes that the organization of public services requires "an approach that recognizes the limits of both markets and bureaucracy, and the need for government as well as management" (257).

A similar message is propounded by Lane (1997). Lane argues persuasively that there are two fundamental aspects to public sector reform: 'proportionality'—economic efficiency and 'impartiality'—linking accountability with fairness. Proportionality refers to the balance between public sector benefits and costs—ie. economic efficiency. Two forms of efficiency are of relevance: internal efficiency and external efficiency. The first—internal efficiency—is concerned with using public sector resources as productively as possible, whereas the second—external efficiency—focuses on the extent to which the public sector achieves its identified goals. Under 'new public management' the efficiency paradigm has taken precedence to such an extent that fairness/equity has become increasingly marginalized as a public sector attribute. In short, new public management with its emphasis on goals and the achievement of objectives has overshadowed public administration with its concern for equity and the rule of law (Lane 1997).

Nonetheless, competition in whatever form, seems to be the key element of modern health care reform. However, with competition comes the need for greater clarity in terms of efficiency, effectiveness, quality of care and consumer/patient satisfaction. The health system and its constituent parts must be accountable for its performance. Accountability and evaluation are interdependent. Both depend upon measurement. A service or system may be evaluated (measured) on the basis of efficiency, effectiveness or some other criteria. However, the results of the evaluation may do no more than highlight what does and what does not work well with the service or system. There may be no compulsion or requirement to address any deficiencies. Accountability, on the other hand, implies that some course of action will follow any evaluative process. Therefore, while it is possible to have evaluation without accountability, one cannot have accountability without evaluation. Thus, the discussion will consider accountability in the health sector, with the understanding that evaluation is an implicit component of the concept.

Accountability

There are many forms of accountability: fiscal accountability; political accountability; professional accountability. These forms of accountability are not mutually exclusive. Indeed, depending on the nature of the provider agency and the health reform environment in which it operates, different forms of accountability are emphasized or stressed. The biggest problem with accountability according to Stein (2001) is that: "Purchasers, providers, consumers and society have different agendas, different goals, and at times, even different values" (172). Therefore, a single concept or measure of accountability will not work because it does not speak to these different agendas, goals and values.

Changes to the structure of the system, primarily through decentralisation and facility designation, have raised questions about whether the existing mechanisms of accountability are still relevant. It is not only the concept of accountability that is under scrutiny but its nature and scope. Flood (1997) in an examination of the reforms in the United Kingdom and New Zealand health systems observes that, "the lines of accountability . . . are too often blurred and there is confusion as to who among central government, purchasers and providers is ultimately responsible." (484).

Accountability issues currently play a prominent role in the debate on Dutch health care. Two aspects—fiscal accountability and professional accountability—are worthy of comment. Fiscal accountability: the government has substantially increased the budgets for health care but in acute care the waiting list seem to be hardly reduced. This raises the question of where the money has gone. Providers must be more transparent in what they do and how they spend their resources. A clear aim of developing the purchasing function in health is to establish a new kind of relationship between purchaser and provider, with a very strong emphasis upon mutual accountability, as in normal market activity. Purchasing means getting something: what do the providers perform for what budget? Professional accountability: this type of accountability is poorly developed in the Netherlands. Little is known publicly about medical failures or more generally the quality of medical performance. However, there is now development towards a greater public transparency in what physicians do and how provider organizations perform.

Stein (2001) suggests that a principle reason for this is the search for one inclusive measure of accountability based on the concept of quality of care. What are required are measures which address the different agendas of the four principals: purchasers, providers, consumers and society. There are four measures which should meet these different agendas: evidence-based outcomes or effectiveness; efficiency used as a means and not an end; performance-based satisfaction, and transparency in reporting (Stein 2001). "Accountability without responsiveness is meaningless, but responsiveness requires willingness by governments—and even more so by citizens—to invest in the improvement of the goods they value most. Accountability is reciprocal and mutually obligating" (186).

Relationships between purchasers and providers are now much more complex and consequently, require far greater maintenance than previously. Perhaps the difficulty lies with treating accountability and autonomy as mutually exclusive concepts. A more useful approach would be to strive for what Loewy (1995) refers to as homeostasis: a balance between autonomy and accountability which not only permits the organization to function efficiently, but also facilitates development and growth.

A fundamental question remains and that is whether or not the new measures introduced as part of the health system reform process do lead to a strengthening or a weakening of the relationship between those responsible for the organization and delivery of services and those who are in receipt of such services. Stewart and Davis (1994) see the reform process as nothing more than an attempt to undermine public accountability by appointing to the boards of public bodies business people who may not understand fully the way in which public service organizations function as they have little or no relevant experience. Thus they will be applying a different and perhaps not wholly appropriate set of criteria to making decisions on priorities and the allocation of resources. Boards comprised of too many people from similar backgrounds will be unbalanced and unrepresentative of the population in the area which they serve. This latter point is one which, all too often, is overlooked; board members are as much servants of the people as they are stewards of public resources. Board members, whether appointed or elected, should reflect broadly the local diversity of population. Moreover, once the board membership is determined, the public should be informed of who is on the board and what their responsibilities are so that local people have some means of holding the board to account for the decisions they take. Failure to do otherwise will only exacerbate further the growing disenchantment among the public with the workings of government at all levels.

Equity and Solidarity

The terms 'equity' and 'solidarity' have particular significance in most public-funded health care systems. Chinitz *et al* (1998) note that systems based on the solidarity principle 'provide redistribution between age groups, between income classes, between single individuals and families and between good and bad health risks'(Chinitz, Preker & Wasem 1998:57). The opening up of public-funded health care systems to competition and other market mechanisms has raised grave concerns about the impact of such developments on equity and solidarity (Gilson 1998). Determining whether or not the concerns have any substance has not been easy. The biggest factor seems to be that there are major methodological constraints that make it very difficult to draw concrete conclusions (Goddard & Smith 2001). Although the evidence is far from conclusive, Chinitz and colleagues (1998) caution countries with universal coverage about embracing uncritically competition measures as the consequences appear not to bode well for equity and solidarity.

Impact of Reform on Hospitals as Provider Organizations

Table 11.1 which presents an overview of the classification of hospitals in some European countries, reveals some interesting patterns. First, not all health care systems have embraced enthusiastically the purchaser-provider split in the hospital sector. For example, Ireland and Greece have retained for the most part centralised control of hospitals. Not all the countries of Central and Eastern Europe have relinquished centralised control of the hospital structure. Albania, Bulgaria, Czech Republic, Estonia and Romania all retain a degree of control whereby hospitals are funded directly by the state. Nonetheless, there has been a clear shift from budgetary provider organisations to more decentralised models with the largest cluster being the autonomous category. The shift to corporatized provider organisations and to wholly privatised providers has not been as great as might be expected. Most governments have only moved one stage, setting up autonomous hospital units in which the authority of the state remains critical. Even in those countries where corporatized and privatised hospitals were created this has not been done on a wholesale basis. Only six countries (Belgium, Italy, Luxembourg, Netherlands, United Kingdom and Lithuania) have extensively embraced the corporatized provider model. In others, some provider units have been transferred but usually not many, and generally those that have, are being carefully monitored (McKee & Healy 2002).

Second, where shifts in provider status have taken place within European health systems there has been less movement among tertiary hospitals than secondary hospital providers. The reason for this is unclear but it may be due to reluctance on the part of governments and the universities affiliated with the tertiary providers to devolve too much authority through turning them into corporatized or privatised entities. Secondary hospitals rarely have this dual form of accountability and this may explain why there has been greater change in this sector with many more countries experimenting with turning some of them into privatised facilities.

Third, the enthusiasm for corporatized and privatised provider organisations is greater among western European countries than it is among central and eastern European countries. Even here the picture is distorted since many western European countries have had private hospitals for several years, some of which were part of the public-funded health system and some were not. This may be in part because of the greater familiarity in western European countries with a market economy and therefore governments are more comfortable with relaxing the constraints on health care provider organisations. The relaxation of the constraints goes only so far however, as few countries, with the possible exception of Bulgaria and Greece, have sanctioned a wholesale change in provider status.

Fourth, perhaps the biggest surprise is how little the crucial elements of health reform (managed markets, competition and the purchase-provider split) have affected provider organisation status in the countries of central and Eastern Europe. Given the emphasis on public sector reform, often as a condition of support from the International Monetary Fund and the World Bank, one would expect to see a greater shift in health care provider organisations from budgetary to corporatised and/or privatised. The table clearly shows that this has not happened. The well-documented concerns about market failure and the political significance of publicly funded health care, may explain why there has been limited movement in European health care systems towards corporatised and/or privatised hospitals. It may also be due to a reluctance of governments and politicians to relinquish control both politically and personally. This observation suggests that more careful and incremental moves to overcome government failures by using market incentives but still avoid the exaggeration of market failures, may be more appropriate. This can create a more balanced use of different public and private means to achieve the diverse (and sometimes) conflicting health policy objectives.

Table 11.1 does not measure changes in the structure of hospital care. Nevertheless, it gives an impression of the scope of hospital reform. The overall picture is that change has been limited. Until now, hospital reform has not led to substantial shifts from the left end of the spectrum with budgetary organisations to the right end of privatised hospitals. Generally speaking, one can observe in all countries political reluctance to large-scale reform. The common pattern seems to be that governments have moved on two stages. Hospitals with the structure of a budgetary organisation were converted into an autonomised hospital; autonomised hospitals, in their turn, were somehow transformed into a corporatised hospital. Governments are still convinced that hospitals should keep public status for reasons of cost control and equity and that privatisation should not be encouraged on the grounds that the behaviour of private hospitals will undermine the public goals of cost control and equity. There is also serious concern about the efficiency of private hospitals relative to public hospitals. In countries with a mixture of public (corporatised) and private hospitals, there seems to be more willingness to privatise hospitals. This is now happening in Germany, for example. In the Netherlands, where all hospitals have a private not-for-profit status, there currently is less resistance to for-profit hospitals than in the recent past.

Conclusion

The two decades leading to the new Millennium were characterized by the 'health reform' movement. Governments of all political persuasions embarked on various modalities of health sector reform in the expectation that the results would produce a better, more efficient and responsive system. The forces driving health care system reform have been concerns over: rising costs, inefficient use of resources, poor quality of care, consumer and provider dissatisfaction and inequity. The approach in most countries to addressing these major concerns has been largely the same, marrying the competition aspects of market by creating separate purchasing and providing functions, with public financing principles to influence the management and organization of health services consumption and provision. Market competition measures were introduced to address the concerns related to consumer and provider satisfaction, in-efficiencies in resource utilization and quality of services delivered. Public finance principles were retained or expanded because they are considered best at cost control and promoting equity through universal access, the solidarity principle.

Works Cited

Chinitz, D., Preker, A., and J. Wasem (1998). Balancing competition and solidarity in health care financing. In *Critical Challenges for Health Care Reform in Europe*, ed. RB Saltman, J Figueras and C Sakellarides. Buckingham: Open University Press.

European Parliament (1998). *Health Care Systems in the EU A Comparative Study*, Directorate General for Research, Working paper, Public Health and Consumer Protection Series SACO 101/rev.EN, Brussels.

Flood, C. M. (1997). Accountability of Health Service Purchasers: Comparing internal markets and managed competition reform models, *Dalhousie Law Journal*, 20(2):470-531.

Gilson, L. (1998). Discussion: In defence and pursuit of equity. *Social Science and Medicine* 47 (12):1891-1896.

Goddard, M. and P. Smith (2001). Equity of access to health services: Theory and evidence from the UK. *Social Science and Medicine* 53:1149-1162.

Hood, C. (1991) A public management for all seasons, *Public Administration*, 69:3-19.

Lane, J-K. (1997). Public sector reform: Only deregulation, privatization and marketization? In *Public Sector Reform, Rationale, trends and problems*, ed. J-K., Lane, London: Sage.

Loewy, E. H. (1995). Of markets, technology, patients and profits, in D. Seedhouse (ed.) *Reforming Health Care, The Philosophy and Practice of International Health Reform*, Wiley,

McKee, M. and J. Healey (eds.) (2002). *Hospitals in a Changing Europe.* European Observatory on Health Care Systems Series, Buckingham: Open University Press

Osborne, R. and T. Gaebler. (1992). *Reinventing Government.* Wokingham: Addison Wesley.

Schneider, M., Cerniauskas, C, and Murauskiene, M. (2000) *Gesundheitssyteme Mittel- und Osteuropas Health Systems of Central and Eastern Europe*, Augsburg: BASYS

Stewart, J. and Davis, H. (1994) A new agenda for local governance, *Public Money and Management*, October-December, pp: 29-36.

Stein, J. G. (2001). *The Cult of Efficiency*, Toronto: Anansi

Walsh, K. (1995). *Public Services and Market Mechanisms—Competition, contacting and the new public management*, Basingstoke/London: Macmillan.

Type Country	**Budgetary**	**Autonomous**	**Corporatised**	**Privatised**
Austria		Y		
Belgium			Y	
Denmark		Y		
Finland		Y		
France		Y		Y
Germany		Y	Y	
Greece	Y	Y		Y
Ireland	Y			
Italy			Y	
Netherlands				Y
Portugal		Y		
Spain		Y		
Sweden		Y		
U.K.			Y	
Albania	Y			
Bulgaria	Y			Y
Czech Republic	Y	Y		
Estonia	Y	Y		
Hungary		Y		
Latvia		Y		
Lithuania		Y	Y	
Macedonia		Y		
Poland		Y		Y
Romania	Y	Y		
Slovak Republic		Y		
Slovenia		Y		

Figure 11.1

Chapter Twelve

Faith-Based Organizations Field Perspectives on Sustainable Heathcare

Chinyelu K. Lee and Dorothy Brewster-Lee

> *The people have the right and duty to participate individually and collectively in the planning and implementation of their health care.*
>
> Alma-Ata, 1978

We have come far from the Alma Ata appeal for health and wholeness requiring the integral participation of communities and targets the development of primary healthcare. The delegates at the 1978 conference envisioned a world with adequate primary healthcare for everyone by the year 2000. Though some achievements have been reached, the ultimate target still eludes us. The re-emergence of the most narrow application of the "absence of disease" as the new definition for health has resulted in preoccupation with short-term, quantitative disease specific goals, the institutionalization of expert agencies, and a decrease in the value given to the building of infrastructure and the empowerment of communities except as tools to achieve exogenously predetermined goals by predetermined mechanisms.

The urgent need to respond to the AIDS crisis, coupled with the American cultural need for instant visible results and impatience with the lack of more tangible result for dollars, were cited as the three primary reasons for this shift from a more holistic development approach. As one Christian Health Association (CHA) stated "AIDS has totally changed the way we do health business in this country." Another health director warned "it's crazy but the hospital administrators just shifted [sexually-transmitted disease] (STD) drug money to [voluntary counseling and] testing (VCT) for the [antiretroviral] (ARV) program. If we are not careful the effort to put all our resources into fighting AIDS may end up creating more fuel for the spread of the disease."

Some recent developments, like time-bounded objectives, are desirable. They are effective tools when applied appropriately. Any process that will curtail the propensity of global leaders to make open-ended promises is a welcomed advance. The United Nation's Declaration of Commitment, with its list of targets dates, is an excellent example of time-bounded objectives being appropriately applied. In 2005, not only will we have a decent idea if 90% of young people (15-24) have access to a variety of things that will reduce their vulnerability to HIV, we will also know how close we are to the target and what areas to focus upon. It is also appropriate to quantify the people who are currently being treated for a specific disease and target future improvements. However, it is when the two indicators above become the only measure of success that they run the risk of steering us away from the ultimate goal of increasing sustainable community centered health delivery for the world's most vulnerable populations.

The development of the expert community has benefited many developing countries with a much higher concentration of technical expertise and "know how." Institutionalization of these expert communities as INGOs presence within developing nations often comes with some drawbacks. International NGOs are accountable to none and represent the desires of no local communities. They often do not come with a framework upon which their excellent pilot projects can be easily or rapidly scaled up and/or turned over to local organizations. Also, many international NGOs do not appear to have a vision of a future where low-income countries are independently managing national health systems without their expertise, so they do not feel the necessity to elaborate an exit plan.

In most sub-Saharan African countries there are faith-based organizations (FBOs) that play an important role in providing healthcare. The points above were articulated during interviews with members of three FBOs responsible for healthcare delivery in Madagascar, Malawi and Kenya. Below, we will describe these organizations and give you specific examples of some of the situations causing them concern. Following these case studies, we will outline some of their basic suggestions for change. It should be noted that while this paper focuses upon cases involving the relationship of Christian FBOs to sustainable development and the international community this is not meant to ignore the work being done by Muslim organizations, which is neither slight nor insignificant. In many countries Christians and Muslims have partnered in organizations like the Malawi Interfaith AIDS Association to create a more effective and integrated response.

Case #1

The first case involves a 354 member Christian health association (CHA), which includes twenty Protestant hospitals, forty Health Centers, forty-two church based health programs and 252 dispensaries in sub-Saharan Africa. Their activities include advocacy, capacity building and technical support. With another FBO in the region, they supply drugs and medical supplies to over 1000 clients including non-profit private and government health facilities in several different countries. Throughout the eighties and nineties this organization received direct funding from European donors to implement programs. They also had individual contracts with donor agencies such as USAID. But, as the head of the organization remarked in Nairobi, "the emergence of HIV/AIDS has completely changed how things are done." There is a new pattern of business resulting from a significant NGO presence and greater emphasis on multi-sector collaborative responses to HIV/AIDS. Collaboration between different sectors is good. However, these consortiums are often heavily populated with and driven by the voice of the experts of international NGOs responding to the demands of donors for achieving predetermined targets and, ostensibly, greater accountability.

In general, INGOs have a management structure that is streamlined and flexible. The Christian health association works with a system of checks and balances. These are key elements in the building of community ownership and subsequent sustainability: attributes that are being devalued by the donor community. Church leadership is considered conservative and the process is slow because of the need to build consensus for high-level decisions in this 354 membership organization.

The head of the CHA recognized the need for efficiency, but noted that "western-style" accountability requires more technical staff in a time when few donors seem interested in funding infrastructure. For Christian health associations establishing competitive salary structures from internal funds creates a tension in the national church where it is difficult to justify paying health officials more than the head of a diocese or synod. The result is that it is more difficult for CHA and member institutions to offer competitive salaries for their professionals, leading to the loss of skilled personnel to INGO expert community. As a result, CHA is less competitive with the international NGOs when it comes to attracting development dollars.

The head of CHA was quick to add that donors and international NGOs are still very interested in using FBOs, just not as lead organizations or true partners. He complained that despite the experience that his organization brought, many of the HIV/AIDS consortiums just wanted to train church workers to follow the program brought from the outside. There is no sense that international NGOs, acting on behalf of donor organizations, are seeking meaningful input from church-based organizations.

A final observation that the head of the CHA shared with us involved the role of the government in healthcare. While it is commonly agreed, in accordance with a number of international agreements, from Alma-Ata to Cotonou, that governments should determine national health strategies, their role as the prime providers of healthcare can conflict with their duty of building effective multi-sector strategies. Due to the extensive network of church-based healthcare, government often sees FBOs as competitors for donor funds rather than resources to be utilized in national health campaigns. This sentiment was echoed by many of the other FBOs that we had the chance to interview. To summarize, the head of the CHA is not opposed to the resources that international NGO and donor agencies are bringing to his country. He is mainly concerned that these groups are not looking for true partnerships with the local organizations, many of them with extensive networks and far more experience providing healthcare in his country. He is concerned that the rules being established by the donor community are placing local organizations at a competitive disadvantage when they are the ones whose capacity must be enhanced for long-term improvements in the quality of healthcare.

Case #2

Our second case study comes from an interview with a representative of a community-based AIDS FBO in southern Africa. Established in 1998, this FBO predates many of the government's District AIDS Coordinating Committee. Until last year it was the only member of civil society doing AIDS work in several of the districts where they have programs. This FBO, which currently operates with one car, one secondhand computer and an administrative budget of $4,100/month for a seventeen member staff, was interested in discussing a recent interaction with an international NGO. After learning, in July 2003, that their country had received money from the Global Fund for their AIDS work, many local FBOs hoped that they would receive funding through their National Council of Churches (NCC) that has been in existence for fifty years and represents more than fifty percent of the population. They were told that this proposal was not feasible because the NCC did not have the capacity to dispense and account for the funds. Local FBOs and NGOs were later told that the list of lead agencies, mostly INGOs, which were selected to manage the funds designated for each district, would be circulated. The FBO in case #2 works in several districts. The decision to manage funds through the districts means that they will have to report to at least five different lead groups. Coordinating and reporting activities to several district lead agencies require resources that no one is interested in providing. Again, we hear the complaint that donors and their international NGO partners are focusing their energies upon funding specific activities: not supporting infrastructure.

Our FBO was familiar with the INGO chosen as the lead agency in the district where the FBO is headquartered. The INGO had initially come to the district to do work on food security issues. However in 2003, the international NGO returned to inform the FBO that they (INGO), a group with little previous experience in AIDS work in the district, were now assuming the role as an umbrella for Global Fund AIDS activities. They stressed that they would be funding activities but not the infrastructure required for supporting those activities. For example, the FBO, as an established organization would only receive money to implement activities on project sites but nothing for the personnel that will be required to supervise and coordinate the activities of these programs. When the FBO complained about this policy, the international NGO threatened to take "their" money elsewhere.[1] The interviewee lamented that the consequence of such restrictions is that many local organizations will bury their administrative costs in order to survive and pay their bills.

This FBO plans to discuss a course of action with members of the church hierarchy that have some access to the country's National AIDS Commission: a luxury shared by few FBOs. However this does not change the fact that all the money for capacity building, at least in this district, is controlled by an international NGO. The FBO official concluded that it was ironic that when the acquisition of Global Funds was first announced, the government said that the money would be used to strengthen local organizations. Unofficial reports suggest that only one of the nineteen lead organizations will be a local organization.[2] These developments do not bode well for building locally controlled sustainable healthcare.

Case #3

This case involves a church based organization that reaches nearly a million youth and young adults via its youth department. Its education department is responsible for 300 primary schools and 152 secondary schools. They also provide Sunday school and scouts activities. In 1974 they established a department specifically responsible for development and technical support. This department has a reforestation program and an extensive program of community health with fifty-seven sites servicing 111 villages. Fifty-two of these sites provide family planning services; forty-five infant-monitoring centers and thirty-four dispensaries. The health services cover a population of 2,267,679 individuals. They have had successful funding relationships with international church groups, international government agencies (USAID), the World Bank and their national government. Most of their health work is self-financed and they have a few income generating development projects.

In 2000, the church organized a multi-department AIDS coordinating committee. The committee has representation from women, laity, education workers, Sunday school workers, scouts, youth and seminary. The coordinating committee has approved church-wide AIDS policies. This committee receives technical support from the successful development department. The committee has targeted their AIDS work at associated congregations and schools located in areas that the government has identified as high-risk zones. Unfortunately, money for administration and coordination by local NGOs and FBOs is hard to come by in this country as well. The traditional European donors have withdrawn their support as a part of a restructuring policy amongst donors in the EU.

This local organization is a vast network with a history of successful development work, yet they are finding it difficult to find money to purchase one computer designated for AIDS work and a salaried employee to coordinate their AIDS program. Few in the donor community seem to recognize what they are bringing to the table as a local organization with demonstrable technical ability. While they have managed to get some funding for other health projects from USAID, they are not being given any money for overhead and administration so they would have to take resources away from their other development projects. They are debating dropping their participation in the INGO managed Child Survival Program because, as stated by a member of the senior team for this organization echoing the sentiments from the cases above, "working with INGOs is difficult, because they don't want to support infrastructure." The unspoken reality is that when INGOs receive grants, the funding designated for administration and infrastructure building is retained at the INGO level.

Discussion

The four concerns which were repeatedly mentioned as indicative of the move away from the more integrated, sustainable, community centered approaches were:

1. preoccupation with short term, quantitative disease specific objectives.
2. the institutionalization of the expert community as a dominating presence of INGOs in Low-Income Countries.
3. overemphasis on the delivery of commodities and devaluation of sustainability which requires the integration of health efforts into communities
4. little recognition of and investment in FBOs as significant community health and development resources

We will elaborate on each of these before moving to the recommendations suggested or implied by the case studies above.

1. Preoccupation with short term, quantitative disease specific objectives

We agree that there is a need for time-bounded specific goals as seen in the Expanded Programme for Immunization (EPI). However, in the UNICEF monograph titled *Community-Based Sustainable Human Development*, Carl Taylor, following the lead of James Grant, writes, "the single most important contribution of the Expanded Programme for Immunization (EPI) to health care generally was demonstration of the power of social mobilization. Social mobilization should start with national leaders but its sustainability depends on continuing community participation." The FBO responders repeatedly complained that every year less money is made available to support the infrastructure of community organizations: capacity building now translates to "don't forget to include some language on training community member in the proposal."

2. The institutionalization of the foreign expert community into the dominating international NGOs presence.

The Second Basic Principle of Sustainable Human Development described in the UNICEF monograph reads, "Action must grow from a combination of a bottom-up and top-down programming. For this to happen in the most informed manner, decision-making needs to be based on a three-way inclusive dialogue involving: The whole community . . . Government & Agency Officials . . . Experts—People with information or useful skills." The strengths of many of the INGOs are their technical expertise and know how. Unfortunately the institutionalization of these organizations often come with drawbacks noted by local organizations. INGOs "*manage*" the resources needed to develop the resources for building local infrastructure.[3] They don't have a framework upon which their excellent pilot projects can be easily or rapidly scaled up. They are accountable to none and represent the desires of no local communities. In the same UNICEF text the authors write "Experts, as outsiders, have an additional role; they can be impartial watchdogs for equity, to warn of intrusions by the powerful." But what happens when the watchdogs become "the powerful?" Who controls them?

Although "local capacity building" is part of every INGOs mission statement there is almost no real incentives to build capacity that would ultimately replace them and so you never hear them speak of their exit plans from a host country. If we were to honestly assess capacity development over the past 15 years, we would probably need to admit that there has been far more international health organizational capacity built up around the DC beltway than in any of the low-income countries. As one Madagasgi FBO health specialist stated, "many of them initially came as either consultants or relief agencies and now they seem to be here to stay. They love our country but they hate us."

3. The perceived overemphasis on the delivery of commodities and devaluation of sustainability, which requires the integration of health efforts into communities.

The local organizations suggested that donor's requirements were aggravating the situation because the donors were only interested in quantifying the number of commodities delivered at the conclusion of a project. The first case study concerning the consensual processing of the CHA outlines several of the problems that donor accountability poses for organizations that are fulfilling their accountability to people in the communities that they are serving.

4. Little recognition of and investment by the international donor community in FBOs as significant community health and development resources.

It was stated that relatively few dollars go to FBO health institutions from international donor agencies. This problem was exacerbated by the restructuring in the European community resulting in a significant decrease in the dollar flow from Europe to African faith institutions. The Global Fund has also failed to provide any significant support to FBOs.[4] They admitted that the problem was not all external; denominations have also failed to fully support their ecumenical health networks, thus diminishing the ability of these networks to capitalize on their united strength. Local FBO leaders also admit that the FBO health community, which delivers up to 40% of the health care in some countries, has not adequately integrated their health activities into their associated faith communities, which some believe could reach more than 90% of the inhabitants in Sub-Saharan Africa.[5] Better integration would greatly increase the capacity of these communities to socially mobilize around health issues.

Recommendations

These cases force us to confront many of the choices that have been made over the past two decades. Perhaps most importantly, *we need to take a closer look at what we mean by "capacity building" and the role of NGOs in this process.* We must ask ourselves, over the past twenty years whose capacity has actually been built for health work? It has not been the national governments and certainly not FBOs, who together represent most of the avenues for truly local, sustainable healthcare in developing nations. We need to take a new look at the role of international NGOs and insist that they arrive with an exit plan. INGOs have an important role to play but their success should be measured in their ability to leave behind locally controlled sustainable projects.

As local organizations with extensive and poignant experience we need to spend more time listening to the voices of FBOs, national medical associations, and other local NGOs. We need to look to them for guidance as we determine the characteristics of the "expert" mechanism that are really helpful. There has been a doubly missed opportunity for building FBOs system for health delivery. First, the FBO community has missed a major opportunity by relying solely upon their departments working health related fields. Thus far the FBO community has failed to empower or mobilized the entire church—by only using a part of their potential resources as the members of Christian Health Associations, the churches themselves, are not supporting their umbrella organizations to the utmost of their abilities. Second, donor agencies, until recently, have not recognized the capacity of the church to be a major player and have also failed to fund them adequately.

We need to reexamine our indicators; continued monitoring of health progress with disease specific indicators and commodities delivered are an important part of the mechanism for monitoring success but not the ONLY serious indicators of success. The recently announced World Health Organization's campaign to treat three million people with ARVs by 2005 is a typical example. Will this campaign aim simply to have three million hands raised at the end of 2005 saying that they have tasted an ARV drug or will it attempt to have a health and social mechanism in place that can insure the on-going long-term delivery of ARVs to three million HIV infected individuals.

The struggle to combat HIV/AIDS must be understood as a part of an ongoing movement to provide adequate basic healthcare to all. FBOs are not just tools to be used by international organizations to achieve these goals, they are local organizations with valuable experience and input necessary for building effective strategies for healthcare in developing nations.

Notes

1. In subsequent follow up interviews other CBOs consistently referred to the INGO as the source of the money that is actually provided by the Global Fund to the National AIDS Commission to be administered by collaboration between District AIDS Coordinating Committees, local government, and the INGO.

2. At the end of the process there were only six lead agencies all of who were INGOs.

3. Case Study #2 provides an example of how this process works.

4. To date, about four percent of the Global Funds allocations went to FBOs according to the December 2003 Progress Report from the Global Fund. http://www.theglobalfund.org/en/files/progressreport.pdf

5. An ambitious figure but not altogether unreasonable when one considers in the Gallup International Millennium survey nearly every single respondent in western Africa and only slightly fewer in eastern and southern Africa claimed that God had high importance in their life. Many of these respondents also told surveyors that they attended religious services at least once a week [Gallup International (2000). *Religion in the world at the end of the millennium*]. Anyone claiming to reach ninety percent of sub-Saharan Africa through religious communities must be including Muslim organizations. The same is true for the response to the question concerning the attending of religious services from the Gallop International question above.

Chapter Thirteen

Malaria and International Health Organizations

Socrates Litsios

Several months ago I received a copy of the evaluation report of the global Roll Back Malaria initiative. Among the many distressing conclusions reached, there was this alarming one: if malaria control is left to governments to plan and execute, malaria will not be controlled.[1]

If true, and I have no reason to question it, we need to ask ourselves how, after a century of international activity, malaria remains the problem that it is. With that question in mind, I started preparing different lists of reasons or factors that have contributed to this state of affairs. These came first to mind:

- Malaria is a complex problem whose manifestation varies from place to place
- Division within the ranks of malariologists concerning how best to approach its control
- 'Dissidents' kept out of important policy discussions

This list brings together several factors which were already present in the early 1920s when the word malariology first officially entered the English language and the League of Nations Malaria Commission was established.

By that time the earlier and very successful efforts to control malaria in the Panama Canal and various commercial enterprises around the world had been dismissed as suitable models to follow, as their costs were prohibitive for any country to afford. Each country had to determine for itself what would work that could be afforded under varying local conditions. It was in that context that various schools of malariology grew.

Different experiences led to differing opinions on how best to proceed. When Lewis Hackett, one of the senior Rockefeller Foundation malariologists and leaders of the American school of malariology, arrived in Italy in 1924, he wrote: "there are so many outstanding malariologists with widely differing theories and mutual jealousies that they cannot be gotten together on any one plan."[2]

These divisions reached a crisis level by the late 1920s when the Rockefeller Foundation malaria experts were at logger heads with the malaria experts of the League of Nations over the issue of the utility of anti-mosquito measures, with the Americans favoring them and the League questioning their value.[3]

The League attempted to control the issue by not inviting individuals such as Ronald Ross, the arch advocate of mosquito control, to participate in its work. But the Foundation was too important an agency to be boycotted since they were funding much of the health work of the league.

Associated with the notion that anti-mosquito measures alone were almost useless was the belief that malaria had to be tackled as part of a wider rural development effort. I've highlighted this particular issue as many believe this to be important today and it was an issue that for various reasons fell by the wayside once WWII started. DDT represented such an advance that it alone might have been sufficient to push it aside.

In any case, so great was DDT's power that the idea of eradicating malaria soon gained the upper hand and so confident were the eradicationists of success that Paul Russell challenged the World Health Assembly in 1955 with, "Whatever WHO decided to do, a campaign for world-wide malaria eradication was already under way."[4]

Not only did WHO decide to join the global campaign, it soon took charge using its Expert Committee to control matters, including stifling opposing voices by not inviting them to participate in its work. Not surprisingly the issue of malaria and rural development did not surface in this context either. This period lasted into the 1970s by which time the goal of eradication had been dropped.

The post-eradication era was characterized by other factors that have not helped the cause of malaria control. It is important to keep in mind that none of the earlier factors that I've mentioned have completely disappeared. These new factors were:

- Malaria never part of mainstream public health
- Premature announcement(s) of vaccine
- Inadequate capacity building efforts
- Corruption

Malaria and main-stream public health have existed in their own, largely distinct, universes since the beginning of the 20th century. Modern public health emerged from countries where malaria was not a major problem. Malaria control emerged in the context of tropical colonialism.

Public health specialists were ignorant of the divisions that split the malaria community, including those divisions that had to do with the meaning of malaria control. For many malariologists the diagnosis and treating of malaria cases was not considered part of a control strategy. Only efforts that aimed at reducing malaria transmission warranted to be included under the heading 'control.'

Being poorly versed in the ins and outs of malaria control, public health specialists fell into a trap in the 1960s and 70s when they accepted the idea that whatever had been achieved by eradication campaigns should be maintained by post-eradication strategies. Maintaining gains required a case detection system that was not geared to improving diagnosis and treatment.

The 1970s were characterized by reduced budgets and very often the return of malaria in epidemic form. With hindsight one can see that at least a decade was lost before it was accepted that diagnosis and treatment were essential components in all malaria control programs. It is during that decade that the promise of a malaria vaccine took hold. It is hard to imagine that this did not influence governments, eager not to invest in a disease whose demise is imminent.

A different variation of this question is why invest in strengthening national capacities to control malaria if the need for new tools is so great as to suggest that little worth while could be achieved with the tools then available. One response is that it is difficult to imagine a health system that can successfully integrate any new tool if it has never had prior experience, even a failed one, in controlling malaria.

The push for vaccines and new drugs contributed to other unfortunate developments—greatly diminished investments in field epidemiological and entomological research, and an increased separation of research from day-to-day control problems. While the WHO malaria control program pushed for health system research which would allow national control programs to experiment with old as well as new tools, nearly all new money went to the search for new tools and the strengthening of national research capacities to participate in these developments. Little was left to invest in control programs.

Training of researchers has developed steadily over the last twenty years or so, while training of nationals responsible for malaria control has not. An unhealthy balance was created where malaria research gained far more prestige than malaria control. Despite recent efforts to support control programs this imbalance still reigns.

Corruption of course saps all efforts, not only those in the field of malaria and public health. Corruption remains very poorly documented but there is no doubt that it is of critical importance. There is a lot of money to be made through the misuse of drugs and insecticides. Combine that with a bit of greed and poor administrative controls and dire consequences are inevitable. Also to be looked at is how donor monies are being misused by some governments who prefer to use these funds to simply create jobs which are often filled by political appointments who totally lack any competence or interest in malaria.

My last list of factors are not really new but they are of growing importance; I am particularly concerned with the bigness of the international machinery that has been put in place over recent years in support of national malaria programs:

- Malaria no longer 'king of diseases' in the tropics and hardly any threat to the industrialized countries of the North
- Malariologists no longer in charge - new breed of 'managers' have taken command
- Basic research dominates and the little applied research that is funded has little operational links with control programs
- More complex international infrastructure; countries need to invest heavily to figure out how to seek funds
- Funding biased towards 'big' projects rather than towards smaller efforts whose aim is steady, long-term growth from 'below'

A couple of years ago I had the opportunity to present some ideas concerning a viable strategy for the future control of malaria.[5] At that time I listed these features:

- Active involvement of communities, especially children, in public projects
- Strengthening of local 'managerial' capacities
- Use of graduate students to support local efforts
- Making 'local' information available to the community
- Strengthening role of international community

Since my topic was the international public health role in the support of such a strategy, I went on to detail what that role might be:

- carry-out demonstration projects aimed at enhancing local capacity to control malaria;
- develop prototype approaches for local capacity building using adaptive management workshop protocols;
- integrate such projects within wider efforts to strengthen national public health functions;
- extrapolate the information needed concerning malaria to other situations of the world and make that information readily and easily available on the web;
- encourage national governments to adopt information policies that are supportive of local public health initiatives;

• encourage and support, technically and financially, national governments in their effort to reform their educational systems to provide needed support to local public health initiatives;
• invite the global applied research community, including historians, to be 'on-call' when specific skills are in short supply, locally and nationally;
• ensure that a representative sample of local initiatives are well evaluated.

Unfortunately, since then I've seen little evidence that these ideas have been picked up by anyone.

To summarize: the failure to control malaria has deep historical roots. Simply throwing money at malaria is not likely to yield results that are sustainable. We need to adopt strategies that have very, very long time horizons, if we are to have any hope of success. Such strategies would invest in local communities and an international infrastructure that is geared to supporting local initiatives. In that context it is absolutely essential that malaria is addressed as part of long-term community development strategies.

Acknowledgements

In preparing this paper I had very useful exchanges with the following specialists: Andy A. Arata, Dilermando Fazito de Rezende, Chev Kidson, Jan Rozendaal, Mike W. Service, Burton H. Singer, and Andy Spielman. Any remaining errors are those of the author and no-one else.

Appendix A
DIVISIONS IN THE 1930'S

Malaria control cannot be dealt with as an isolated problem separate from other social, medical and public health affairs. (S. Price James, leading member of League's malaria committee)[6]

Malaria is a health and social problem; it must be attacked simultaneously from both these angles. (Selskar Gunn, Rockefeller Foundation Vice-President)[7]

* * * * *

[T]he causes of malaria . . . are in the main independent of the ignorance and poverty of its victims and can be separately handled. (Lewis Hackett, Rockefeller Foundation malariologist)[8]

In those areas where malaria is the outstanding social and health problem, the resources of the health administration, specially augmented where necessary, should be directed chiefly towards malaria control, even if this should entail the restriction of other public health activities, until malaria is no longer of major importance. (Paul Russell, Rockefeller Foundation malariologist)[9]

Notes

1. *Achieving Impact: Roll Back Malaria in the Next Phase—Final Report of the External Evaluation of Roll Back Malaria* (London and Liverpool Schools of Tropical Medicine) August 29, 2002.

2. Quoted in Darwin H. Stapleton, "Internationalism and nationalism, the Rockefeller Foundation, public health and malaria in Italy, 1923-1951," *Parassitologia* 42 (June 2000): 127-134.

3. For examples of these divisions see Appendix A above. See also Chapter 4 in this book.

4. World Health Organization, Eighth World Health Assembly, Mexico 10-27 May 1955. Official Records of the WHO No. 63: 205.

5. Socrates Litsios, "Malaria Control and the Future of International Public Health," in Elizabeth Casman and Hadi Dowlatabadi ed., *The Contextual Determinants of Malaria* (Washington DC: Resources for the Future, 2002)

6. "Principles and Methods of Antimalarial Measures in Europe," Second General Report of the Malaria Commission (Geneva: League of Nations Health Organization, 1927) prepared by S. P. James.

7. League of Nations, Report of the Intergovernmental Conference on Far-Eastern Countries on Rural Hygiene, Bandoeng, August 3rd to 13th (Geneva: League of Nations, 1937), taken from Introduction written by Gunn.

8. Lewis W. Hackett, *Malaria in Europe* (Oxford: Oxford University Press, 1937).

9. *League of Nations, Report of the Intergovernmental Conference on Far-Eastern Countries on Rural Hygiene, Bandoeng, August 3rd to 13th* (Geneva: League of Nations, 1937). Taken from Malaria committee report written by Russell.

Chapter Fourteen

Obligations and Accountability in International Public Health: State and Non-State Actors

Stephen R. Latham

When bioethicists speak about the ethics of medical interventions, they tend to stress two main ideas: respect for patient autonomy, and patients' rights. The notion of "respect for autonomy," of course, is among the most famous of bioethics buzz-phrases. This originally Kantian notion was prominently introduced into the bioethics field by the Belmont Report on research on human subjects[1] in the 1970s, and was firmly established as one of four central principles of bioethics by the leading text in American bioethics[2]. Even among those four principles—autonomy, non-maleficence, beneficence, and justice (the so-called "Georgetown mantra")—the principle of respect for autonomy has been pushed to prominence[3]. The joke is that when a bioethicist goes to remember the four principles, he says, "Well, let's see. There's autonomy, autonomy . . . um, autonomy . . . and I forget the other one!"

Discussion of "bedside bioethics" has focused largely on autonomy conceived of as patient choice from among a selection of options offered by the system. Roughly speaking, one is thought to be enhancing patient autonomy if one describes to the patient, in language he or she can understand, the array of medical choices available—including the choice not to undergo any medical intervention—and then allows the patient to make the selection according to his or her own values and priorities. Note that this "choice" view of autonomy is not the Kantian notion of autonomy—it says nothing, for example, about the proper grounds for autonomous choices.[4]

Autonomy, though, plays a much less decisive role in bioethicists' discussions about public health. Historically, the ethical discourse of public health has stressed considerations of justice and distribution over those of autonomy; this is largely because public health involves analysis at the level of populations rather than of (autonomous) individuals. There has been a great deal of serious and excellent work done on justice and distribution in the public health arena in the last decades. More recently, though, the public health ethics debate has been hijacked by world events. Take the current controversy over Georgetown Professor Lawrence Gostin's Model State Emergency Health Powers Act, for example: the Model Act has been widely criticized for embodying a state-centered command-and-control model of public health, with less local participation and information than its critics desire.[5] That debate, obviously, has been informed by the events of 9/11, by the anthrax scare, by the sudden outbreak of SARS and by the threatened outbreak of human-to-human avian influenza. Where earlier we thought of "public health" as disease eradication, and worried about the justice of the distribution of our efforts and resources toward that goal, we are increasingly tempted to think of it as crisis—and terror—management. Much of the contemporary public health ethics agenda is about the proper way to conduct sudden interventions to take care of bioterror attacks—or perhaps to take care of people in the wake of floods or pandemics. That kind of crisis management often requires states to use their coercive power: to quarantine, to evacuate threatened regions, to inoculate. The national bioethics debate in public health has focused on how and where to set the limits of that state power.

States, classically, have a monopoly on the use of force within their borders—that, in fact, was Max Weber's definition of the modern state.[6] But of course states do not have a monopoly on resources expended in public health, and traditionally don't have much power to coerce health-related behaviors beyond their own borders. This means that some of the post-9/11 public-health ethics vocabulary—those new worries about the proper scope of and limits to state coercive power—don't apply particularly well in the international public health arena. Nonetheless, the events of 9/11—and, more recently, the relentless march across the globe of avian influenza—have had an effect on our attitudes toward the role of states in international public health. Again, the vocabulary of disease eradication has been partially displaced by the vocabulary of crisis management. Something new has been added to our list of motivations for offering public health assistance abroad: generosity is still a motive; and self-interest is still a motive (as when aid is given in return for keeping markets open). But public-health aid is now seen as closely tied to national security interests.[7] States need to eradicate disease not just for the sake of the people involved, but to prevent important geopolitical regions from sinking into disorder and war. Moreover, they need to make public health interventions in order to stop diseases from spreading across their own borders; they need, in other words, to fight the diseases abroad so that they don't have to fight them at home.

We still lack a well-developed vocabulary with which to discuss the ethical constraints on such self-interested public health interventions by states; we lack, in other words, a decent political theory of public health. (The philosopher Onora O'Neill has written an excellent paper about this, to which this chapter is heavily indebted.)[8] Could we perhaps advance the discussion by starting from the other end of the international public health problem, and addressing ourselves not to state's interests, but to human rights?

Unfortunately not. The problem with rights discourse—as O'Neill points out—is that rights imply correlative duties. Think about a classic human right, e.g., the right not to be tortured. Such a right involves—indeed, is defined by—a correlative duty imposed upon all others to refrain from torturing. This is the structure of most so-called "negative rights"—rights to be free from coercion or fraud, rights not to be imprisoned without charge, rights not to be tortured and so on. They depend upon correlative obligations, shared by everyone in the world, to refrain from doing various evil things. But there are other kinds of rights—"positive rights"—that seem to have a different structure. Rather than simply imposing obligations on everyone to refrain from doing evil, they seem to impose duties upon particular people affirmatively to go forth and do good. Rights to due process can be like this, and of course rights to health care are like this. If a person has a right to health care, it's not enough for all of us simply to refrain from interfering in that person's life; instead, someone with the relevant skills and resources has to go out and supply the person with the healthcare to which they are entitled by right.

Who has the obligation to be that supplier? Normally, we think of states as having that obligation with regard to their own citizens; but there is no history of regarding states as bearers of duties toward the citizens of other states. Indeed, the parable of the Good Samaritan notwithstanding, our tradition of recognizing and enforcing moral and legal duties to assist others—particularly if those others are strangers—is notably thin. If a person is not in a pre-existing fiduciary relationship with us (as a child is with a parent, or a patient with a doctor), we generally have no personal duty to assist her. Similarly, states generally have no duties to assist those who aren't their own citizens.

An important exception to this general rule is triggered if it is our own actions that have put the other person in harm's way. In those circumstances, we do acquire a "duty to rescue"—to remove people from the danger to which we've exposed them. I normally have no duty to leap into the pond to save the drowning stranger, but if the stranger is drowning because I knocked him into the pond, I do have such a duty. According to political theorist Thomas Pogge[9], wealthy states owe aid to the poor and sick of the world because they are responsible for the rise and continued dominance of an international economic system that created and sustains massive poverty worldwide. One might argue—though Pogge does not argue the point in just this way—that wealthy states have a "duty to rescue" those whose lives their own actions have put in danger. The point, of course, is deeply contentious and will not supply us, any day soon, with a shared understanding of the basis for state responsibilities for global public health.

If the political theory of states' involvement in global public health appears to be in a muddle, how much greater is the muddle when we turn to consider non-state actors: transnational corporations, foundations, NGOs, international institutions? These are among the most active agents in the international public health field, yet we have little vocabulary with which to discuss their motivations, obligations, aims, or the ethical constraints to which their activities ought to be subject.

Questions of motivation and obligation are of primary importance here. If it turns out that the public-health activities of non-state actors are based in self-interest, then political-theory arguments about the legitimacy of, and limits to, the pursuit of self-interest will apply to those actors as well as to states. If their pursuits are purely gratuitous, it will likely be inappropriate to make many demands upon them regarding their aims or their governance; the most we might legitimately ask of them is that they avoid harming others with their purely voluntary activities. If, on the other hand, they act neither out of self-interest nor purely gratuitously, but under force of obligation, then there may be some ethical constraints on the ways in which their obligations may be met.

Of course, not all non-state actors are in the same conceptual boat with regard to these problems. Like states, private corporations may sometimes engage in public-health efforts for the sake of advancing their interests elsewhere; but such motivations cannot generally be ascribed to foundations, NGOs or international health organizations. Non-state organizations generally have little coercive power to worry about limiting—but some private corporations have indeed controlled local military or police power from time to time, and some international organizations are backed by states' military power. Some international aid organizations, and some private corporations, might be implicated in the maintenance of a harmful international economic scheme, and might therefore have acquired duties to rescue those they've put in peril—but this hardly describes the situation of most foundations or NGOs.

Let us assume that the public-health activities of NGOs and foundations are not based in self-interest, and that they have no obligations based on Pogge-style restorative justice or a duty to rescue those they've put in peril. Are their efforts then purely gratuitous, or are they somehow grounded in another kind of obligation?

There is another method by which we might generate an obligation of public-health assistance that could run to private parties in the international health context, including to NGOs and foundations. In the early 1970s a team of business-ethics writers came up with what they called "The Kew Gardens Principles" to explain when a private party—whether a person or a firm—might acquire a duty to assist a stranger.[10] (The name "Kew Gardens Principles" was a reference to the famous murder of Kitty Genovese in the Kew Gardens apartments in New York; thirty-eight people witnessed her slow murder and did nothing.) They concluded that 1) where a person is in critical need; and 2) a party is sufficiently proximate to the person in need to be aware of the need and to be able to act to address it; and where 3) the proximate party is actually capable of assisting; and 4) where assistance from others is not forthcoming, then the party has a positive moral duty to help the person in need. The theme here is basically that, in the proper circumstances, a great need can give rise to obligations in others. We can easily see how this might work in the international public health context, where great needs are obvious everywhere; where private parties including foundations and NGOs are aware of, and have the ability to assist in addressing, those needs; and where assistance from other actors—local governments, for example—is not forthcoming.

So, we can think of ways in which private organizations might acquire obligations to assist in international public health. Does this imply any ethical constraints on the assistance they give? I would argue that it does. A purely gratuitous gift may be given anonymously. The gift itself may be anything (or at least anything that is not harmful). There need be no explanation offered of the choice of gift, or of why it was given. But assistance given as a result of obligation to aid the recipient comes with some ethical burdens. For one thing, the recipient must be given something that actually does aid him. Practical assurance of this fact, especially in an area as complicated and subtle as public-health intervention, is best secured by giving the recipient some voice in determining how he is best aided: what his needs are, and how they can best be addressed. He must also be told to whom he can complain when the assistance given is a mismatch for the problem at hand, or starts to give rise to unforeseen and unpleasant consequences—and he must have some practical assurance that his complaints will be taken seriously. In this not-very-roundabout way, any obligation to render complex assistance to others gives rise to collateral obligations of accountability and good governance.

It is undoubtedly a great advantage that NGOs and foundations can pursue public health problems across borders fairly freely, without (much) constraint by state interests or broader political conflict. A distinct disadvantage, though, is that most of these organizations are not democratically governed, and don't have built-in accountability mechanisms. Their governance is not particularly transparent. Even those that are of tremendous goodwill are only transparent to the extent that they deliberately choose to be so—and many of them have decided not to be.

There's a worry in the administrative law world that accountability, transparency and democratic process are vanishing in two different directions[11]. On the one hand, decisions are being made by non-elected global bodies—sometimes firms, sometimes global organizations like the World Trade Organization ("WTO"). Think, for example, about the recent headline issue of steel tariffs.[12] If you don't like what the WTO has done to eliminate United States steel tariffs, whom do you vote against? There is no one. (If you didn't like the original tariffs themselves, of course, you could have voted against President Bush.) Again, if you don't like the way we're doing agricultural subsidies, you can vote against Bush; but if they're abolished by the WTO there is no one to hold politically accountable. There are a number of decisions that are disappearing up into "unaccountability" at the global level, because they are being made within bodies that are neither democratically responsive nor transparent. On the other hand, another set of decisions is being pushed down into "unaccountability" at the local level. During the conference on which this book is based, for example, Carl Taylor mentioned approvingly the idea of having Institutional Review Boards ("IRBs") for a number of different functions beyond research on human subjects. But the American IRB is, in fact, one example of a mechanism by which accountability and transparency are made to disappear in the "downward" direction. The government is increasingly involved in negotiating the content of regulations with the industries being regulated; and with passing regulations which, instead of specifying what's allowed to go forward, specify a process by which affected parties will determine for themselves what's allowed to go forward, with minimal mechanisms for central accounting, for uniformity, for transparency. The IRB is precisely the latter sort of mechanism. Most people who enroll in a study never hear that an IRB looked at the study, have no idea who the IRB members are or what standards they operate under, and have never had an opportunity to help choose who the members of an IRB are. And IRBs don't report to one another or work with one another to set up uniform standards. Transparency and accountability vanish into a form of local governance for which no elected person or body is directly responsible.

Thus we have transnational decision makers on the one hand and local, voluntary, participatory bodies on the other, and though each has a lot of virtues, neither is particularly accountable, transparent or representative. I am tempted, now, to map these brief comments onto the addresses made by the keynote speakers for the conference on which this volume is based. Peter Goldmark, director of the Global and Regional Air Program for Environmental Defense, remarked on the transnational flexibility, responsiveness and inventiveness of international foundations, but also remarked pointedly that the accountability of those foundations is not—and should not be—to any oversight body, but only "to the health of our children." Carl Taylor of Johns Hopkins's Bloomberg School of Public Health several times made remarks about starting his locally-based, participatory public health team projects "before the government noticed," and about "coming in under the government's radar screen." These remarks might stand, roughly, for the worry that, thanks to the involvement of NGOs and foundations in global public health, important public health programs are being implemented both above the level of democratically accountable government, and below it.

To conclude: foundations and NGOs are among the important international actors who, on one philosophical view, might have a duty, grounded in their ability and understanding and resources, to help those in need of public health assistance. But in many cases their approaches seem to lack important political features: visibility of the organizational structure, accountability to the people affected by their decisions—features that we normally associate with a healthy democratic process, and features that are ethically required to assure that those actors are meeting their obligations to those in need. Whom can the needy blame if a project doesn't come out right? Who's in charge here? Who's responsible for having set things up this way? These are important questions—some of the most important questions of political theory. And in the area of international public health, answers to these important questions will not be forthcoming as long as foundations and NGOs think of themselves merely as gratuitous givers of aid, and therefore free simply to fly above politics, governance and accountability, or to sneak in below them.

Notes

1. National Commission for the Protection of Human Subjects of Biomedical and Behavioural Research, "The Belmont Report: Ethical Principles and Guidelines for the Protection of Human Subjects of Research." April 18, 1979.

2. Beauchamp & Childress, *Principles of Biomedical Ethics* 5th ed. (New York: Oxford University Press, 2001).

3. On the comparative importance of the principle of autonomy in American bioethics, *see* Paul Root Wolpe, "The Triumph of Autonomy in American Bioethics: A Sociological View," in Raymond DeVries and Janardan Subedi, eds., *Bioethics and Society: Constructing the Ethical Enterprise* (New York: Prentice-Hall, 1998).

4. On the many non-Kantian senses of "autonomy" in bioethics, see Baroness Onora O'Neill's Gifford Lectures, *Autonomy and Trust in Bioethics* (Cambridge University Press, 2001), chs. 2 and 4; a similar but more detailed account is in her *Bounds of Justice* (Cambridge University Press, 2000), ch. 2.

5. See, e.g., George J. Annas, Bioterrorism, "Public Health and Civil Liberties," 17 *New England Journal of Medicine* 346 (April 25, 2002):1337-1342.

6. The definition is in Weber's widely-anthologized essay, "Politics as a Vocation."

7. See David P. Fidler, "Public Health and National Security in the Global Age: Infectious Diseases, Bioterrorism, and Realpolitik," *George Washington International Law Review* 35 (2003): 787

8. Onora O'Neill, "Public Health or Clinical Ethics: Thinking Beyond Borders," *Ethics & International Affairs* 16 (2002): 2, 35-45

9. See Thomas Pogge, "World Poverty and Human Rights," *Ethics and International Affairs* 19 (March 2005): 1-7. Pogge does not frame his argument in terms of duty-to-rescue.

10. John G. Simon, Charles W. Powers and Jon P. Gunnemann, "The Responsibilities of Corporations and Their Owners," in *The Ethical Investor: Universities and Corporate Responsibility* (New Haven: Yale University Press, 1972).

11. See, e.g., Martin Shapiro, "Administrative Law Unbounded: Reflections on Government and Governance," *Indiana Journal of Global Legal Studies* 8 (Spring 2001): 369-77.

12. See Mike Allen, "President to Drop Tariffs on Steel; Bush Seeks to Avoid A Trade War and Its Political Fallout," *Washington Post*, Monday, Dec. 1, 2003, p. A01.

Part Three

Toward Tomorrow

Chapter Fifteen

Racing Against the Global Industrial Hangover—Thoughts on Surviving the Morning After

Peter C. Goldmark, Jr.

I'm going to talk from the heart this morning. I am not going to be cautious or careful. I am an American who has recently come home after living five years abroad. From abroad, I must tell you that it looks to most people as if America has lost her way. As if the country that was the pole star for many people throughout the world for a large part of this century had gotten confused and wandered off course, almost without realizing it herself. There is resentment. But it is not the resentment that hurts. It is the pity that hurts, and the genuine sadness at our confusion. That is what is painful.

And yet—I come home, and I discover what I knew before: that there are more people of good will, of determination, of ingenuity and compassion in any square mile of American territory than you could possibly imagine. So let us come together for this conference with a sense, humbly conceived but daringly imagined, that in this room we harbor in our midst some of those people, and in our hearts some of that hope for what American can be at her best. I'm going to talk about four things this morning that begin with the letter "C." And they all relate to health.

I'm going to talk about *currencies.*

I'm going to talk about the "*crunch.*"

I'm going to talk about our children and our *children's children.*

And I'm going to talk about the *critical* role of foundations.

I think it is useful to assess what foundations have accomplished in the past in terms of medicine and health . . . luckily there are many here who know more about that than I do. Yes—I believe that you can and must learn from the past. But I also believe that what we must do now sounds at once trite—and daunting: and that is to build a new framework for looking at the future because the old frameworks that served us well in the past have come to an end of their usefulness. And building new intellectual frameworks, new mindscapes is real tough for us human beings. Generally we like it comfortable, and we like things the way they are. We have to be bludgeoned and pummeled before we will abandon an old framework and create a new one.

Where are we?

We remember storms. But we rarely remember what it was like just before the storm. We live today in a period before a storm—*before* a period of searing difficulty, *after* a period of enormous accumulation of wealth, power and technology in the North, *during* the decline of one international order and the search for a new one. Two plays set before World War I and the Bolshevik revolution recall another such uneasy, hauntingly disturbed moment before a storm: *Heartbreak House*, by George Bernard Shaw, and *The Cherry Orchard*, by Anton Chekhov. I am drawn to that moment—the period before World War I—and those plays because they capture something that is present once more: a foreboding that deep forces we can barely discern are reshaping the world we knew. And in the face of this unsettling, often reluctant, recognition, we are today at once appalled and diverted—as we were at the beginning of the twentieth century—by great wealth, by intense artistic exploration, by dazzling technological breakthroughs, by self-indulgent narcissism, and by a feeling of helplessness in the public arena.

We share with that pre-World War I period a dwindling confidence; we see glowering storm clouds on the horizon; we experience the psychological impossibility of living every minute with our consciousness acutely tuned to impending danger; and with comfortable attachment to the last beats and chords of the known rhythms of a way of life now substantially undermined. Like Chekhov, I can hear the thwacks of the axes as they chop down the trees in the cherry orchard. Can you?

When will the deep fault lines shift? We cannot know. But we have felt already some of the advance tremors. We sense, fleetingly, that a new world is being prepared. But as long as the old record plays, it seems that we know how to dance only to the melody to which we have grown accustomed.

The Ultimata

The human community faces two ultimata. To address them should constitute an important agenda for philanthropy over the next two decades. "Ultimatum" is a strong and threatening word. Why is it appropriate? Because the threats we face are global and immense—literally capable of changing the outcomes of the human adventure on this planet.

The first ultimatum is the threat of weapons of mass destruction. This threat presents itself today in two forms: catastrophic terrorism, and use of weapons of mass destruction (WMD) by one state against another. In these forms, and in others that may later arise, this threat is permanent. As long as there survives on this planet a substantial community of humans who tinker with technology, who explore, exploit and engineer in the realm of science, then the capacity for and instruments of mass destruction will be present as well.

The second ultimatum is the progressive destabilization of the global environment. This danger is more gradual than the first, but it is relentlessly cumulative. Like tooth decay, the longer it builds, the more difficult and expensive it becomes to reverse. The possibility of progressive environmental degradation is inherent in the particular developmental path that the pattern of human industrial activity on this planet has taken over the last few hundred years. Now the time for this model of headlong exploitation of natural resources is running out. The earth is our host and we are its guests, and we are on the verge of destroying the only habitat we will ever have.

For the human community to survive in anything like the condition to which we aspire, these two ultimata must be addressed successfully. But there is also a third dimension of the contemporary situation that presents a moral and pragmatic imperative for action, and that is tightly linked to the two ultimata: this is the large and growing disparity in wealth and circumstance between the richest and poorest among us.

To understand how these issues connect and to develop an approach to all three together that can command a high degree of allegiance around the globe is the biggest challenge we have faced during our existence as a species. It is a critical challenge for philanthropy. Can we get it, together?

The Dilemma

Here is the dilemma, stripped down to its ugly but simplest elements: More states and terrorist groups are getting access to weapons of mass destruction. These weapons grow smaller, cheaper, easier to get, and more destructive. This is an immense threat, and one—even with 9/11 and suicide bombings in our headlines—that is underestimated. But where, exactly, do the roots of terrorism intersect the reality of poverty and the gap between the rich and the poor?

We live in a world where the rich have been getting a lot richer, and where the poor in many countries have been getting poorer, more numerous, and more desperate. And where for many any reasonable prospect of opportunity or improvement in the future is absent. And that is key . . . not only actual self-improvement, but the *prospects* for it. Poverty and desperation do not directly cause terrorism. Most terrorists do not ride forth from the poorest slums of the world . . . but the clamp of despair, the fetid hothouse of poverty that traps so many, create an environment in which many tolerate or even tacitly support the use of violence against those who hold the keys to the storehouse.

Consider the case of the suicide terrorist. The mission of the suicide terrorist succeeds only if his work is applauded and celebrated, whether publicly or in private, by those who are called upon to support his cause. Ten young men and women will step forward to take his place only if he dies a martyr and a hero. If he dies an obscure death, or is considered stupid to have thrown away his own life in an unjustified act, no one will step forward to take his place. Suicide terrorism requires a "breeder chain reaction" of faith and imitation.

There is no way to prevent completely all terrorism with weapons of mass destruction. The task is to marginalize catastrophic terrorism in the minds and hearts of human beings, and to make it as lonely, as difficult, as unrewarding and as despised as possible. This requires giving people a stake in the global order that will lead them to condemn rather than celebrate acts of mass violence and terrorism. Today billions in the South do not have a stake in the global order, such as it is, and hundreds of millions in the North are ambivalent about whether they really desire a global framework of fairness and the global institutions this would imply.

And all of us exist today in an order whose patterns of economic production and consumption will be relatively short-lived.

The period of deepening environmental deterioration does not lie in the future. It has begun. There are no more cod in the Grand Banks fishery. There is less fresh water, of lower quality, available in the world for household use and irrigation. Water tables beneath the three largest grain-producing plains [1] and elsewhere are sinking. The world's climate is warming dangerously and threatens to curtail agricultural productivity, affect the availability of fresh water, and trigger the spread of tropical diseases, the destruction of low-level coastlands, and the death of many coral reefs, among other effects. Tropical forests continue to be cut and burned away every year in massive chunks.

So the ugly truth here is that the pattern of economic consumption and production in which we live today will not work much longer either for the one billion in the developed world or for the five billion who are either mired in poverty or racing frantically to join the ranks of the developed world.

The developed world they are rushing to join is the icon of the old order, but the icon is starting to crumble. And to minimize and marginalize the people who would wreak havoc with terrible weapons, the global order must be fair, it must be inclusive—and it must be sustainable.

What does such an order look like? Are we having that debate in our country? Not really. It is our job to have that debate and to then to seek to build that order.

From Terrorism to Health

So here is how we get from terrorism to health.

People who have a stake in the system do not constitute a welcoming surrounding for terrorism. And most terrorist groups cannot live easily outside of an atmosphere that glorifies them or at least winks at their violence. Look at the difference between Al-Qaeda and the Aryan Nation in our own country. Both are apostles of violence. But one is admired, the other is scorned. One is virulent, the other is marginal. One is a big threat—the other is an annoying sideshow. To marginalize terrorists, there must be opportunity and hope for those they pretend to represent—not because the poor are terrorists, but because without the neutrality or the complicity of the desperate, the poor and the aggrieved, the terrorist is isolated, both literally and ideologically.

And what kind of opportunity might the poor want?

Well . . . which is worth more to you—your health, or your money? Some of you may hesitate. For you and me, our money is in fact one avenue for acquiring the support for the health you need. But you probably value health a lot —a whole lot. And those of you who have had a heart attack, or lost a child to disease, may value your health and that of your family more than all your money.

Now let me change the question so that it pinches a little more. If you were a wealthy person and lived in 1910 in Europe, let's say —just before World War I—which would you value more: your health, or your Tsarist bonds?

Why do I put it that way?

Because currencies matter. And we must define what is the currency of opportunity.

What is money for? It's not the paper that has value, it's what it can buy. And what it can buy depends upon it being part of a stable monetary system, and beyond that upon being part of a stable and sustainable economic system for the production and consumption of goods and services. In such a system, one thing most people want to buy around the world, and that includes most particularly the poor who don't have it, is good health for their families and themselves.

Let's translate this another way: one of the things the rich have access to is health, and one of the things the poor most want—one of the things they are "poor in"—is access to health. *Health lies is right at the center of the dilemma about human equity on this planet.* Health services are one of the most valuable, perhaps *the* most valuable, currency we have. Sometimes we lose sight of this. Health is one of the most important things people want development and opportunity and income and improvement *for.* Any attempt to imagine a global order of fairness rather than despair, of sustainability rather than environmental collapse, will place health right at the center of its value system.

But because of the ultimata—the spread of weapons of mass destruction and the accelerating deterioration of thc cnvironment, all our present dollars and euros and yen are increasingly looking like Tsarist bonds. We have entered the period of thc "crunch". Don't think that because you're well off you have nothing to worry about. If you're on the wrong train, it doesn't matter how much you paid for your ticket.

A Basic Health Package

Let us conduct what Einstein used to call a "thought experiment". Let us imagine we had put in place around the world a basic health package. This package would consist of:

1. A universal minimum level of immunization, prenatal care, and family planning.
2. A universal minimum level, adapted locally, of community-based measures to provide clean water, basic sewage, and other simple public health elements.
3. A system of simple, community-level primary care with emphasis on self-care at the household level, such as was achieved in Bangladesh with ORT therapy.

4. A sharply beefed-up worldwide network of disease surveillance and prevention centers.

First of all, what would that cost? Perhaps $100 billion per year. Not much, when measured by the standard of wars, luxury automobiles, expenditures on cosmetics and plastic surgery, etc. Not much by the standard of tax breaks for the rich or subsidies to large agribusinesses that are already profitable. Not much by the standards of tax subsidies to the fossil fuel industries who contribute the bulk of the gases that are over-heating our climate.

And what would it mean?

A global order that can promise your child basic health care is one you would want to support. A global order that helped your family resist and prevent disease is one you might defend against those who would destroy it.

Health is the fundamental currency of the equity dilemma—the "fairness imperative", if you prefer. It is the currency of hope for your children.

But we still have that other dimension of the overall dilemma to address. The new environmental realities tell us that we have entered the period of the "crunch". So the workable road forward does not consist of helping the poor achieve improved economic status within the present system that obtains in the north and that the galloping developing countries like China are seeking so assiduously to imitate. We have, at a guess, two decades in which to reorient our economic system.

We will remain yoked to technology. Technology is one of the forces that got us into the crunch, and it is one of the resources we will need to get us out of it. We will need to continue producing lots of energy—but not the kind we have now. Growing enough food for 8 billion people and defending against new or souped up old diseases will require science and technology. Transportation will require technology—but not the kind we have now. We cannot just shrug our shoulders and walk away from the great Faustian bargain with Science we entered into so enthusiastically and cavalierly. There may have been a moment when that was possible—but it was probably before the beginning of the twentieth century, and we did not recognize that moment or that choice. It is too late now. We will need more, and better, and more appropriate technology, applied on behalf of a sustainable pattern of existence rather than harnessed to accelerating exploitation under the existing model.

We will need a concept of minimum standards for everyone on this globe. This is not a daring new idea. Every developed country on the globe has its own minimum standards—of health, of income, of education, of welfare generally. Now we have to conceive of that on a global level. And the health element should be the first. It is far from the most expensive. It is probably the most powerful, in terms of its positive effect on people's lives and in terms of the intensity with which people want it. And not unimportantly, it is probably the global minimum standard that can be most easily understood and accepted by the rich countries of the North.

Now—let's turn to the other two words that bring us here today – as they apply to a sector of our society from whom we rightfully expect, and need, bold nad far-seeing action to address these problems.

Accountability

Should foundations be accountable to any agency external to themselves? Of course not. Look: foundations are vehicles of personal wealth, whether singular or pooled, and the yeast, the individualism, the entrepreneurial initiative and risk-taking of the philanthropic sector depend upon being free to take on whatever they choose in whatever way they choose. There is an entire debate, as you know, as to whether foundations should be more accountable to their founders—their original endowers. That debate is a sterile and politicized one. Founders have a range of tools available to them to circumscribe, or not, as they choose, what the trustees of a foundation do. But no one's reach beyond the grave is perfect – and maybe it is better that way.

A large part of the value of foundations in our society derives from three characteristics:

1. Foundations can take on very difficult and very controversial projects. They are at the far end of the spectrum of organizations in that regard. That is of enormous value. George Soros can go in and set up universities all over eastern Europe as those countries were reborn. Bill Gates can tackle global health for the poor. MacArthur can set up the genius program, which raised a great many eyebrows at the beginning. Rockefeller can fund population programs that outrage the right wing conservatives. And yes: a Texas oilman can set up a foundation to study how the Indian rope trick is done and whether aliens are kidnapping people and putting plastic implants in their ears.

To whom would we make foundations accountable on these kinds of issues? Some arm of government? Some super board of overseers or trustees? Everyone can comment and throw stones at what foundations do, and foundations whose activities are not transparent should be criticized and even penalized. But set up some outside governance structure to pass on their programs and initiatives? Are we crazy?

2. The second virtue of foundations is that they can take the long view -- longer than other sectors of our society. That is crucial now because so many of our problems have time horizons expressed in decades—far longer than the working time horizons of governments or of private corporations. It's a wonderful thing that the Ford Foundation started supporting Community Development three decades ago and stuck with it over the long haul. It's a valuable thing that Carnegie started working with Russia four decades ago and stuck with it.

3. The third irreplaceable characteristic of foundations is that they are hard to intimidate. They may be slow, they may sometimes be bureaucratic or worse—but they sure are hard to intimidate. Foundations and the independent press are the sectors of our society whose values and constitution make them most immune to political influence or commercial pressure. In the American system we forget sometimes just how thin is the line that separates the pluralism

and freedom that Americans enjoy from the kind of suffocating autocracy that marks, for instance, Russia today. Foundations form an important bulwark in strengthening that thin line—the line that is always, always under attack.

So for accountability, let's stick with the Internal Revenue Service check on whether the purposes and activities are charitable, and whether the expenditures and other items on their tick-list are properly done. Let the accountability test for what foundations do substantively remain in the give and take of ideas in the public arena. I challenge anyone here to propose a formal structure of outside accountability and review that would not run in the direction of restraining, limiting, making foundations more cautious and skittish.

Equity

Yes. Equity is important—in many senses.

For the first time we must conceive and work toward a scheme of minimum equity on a global scale. By that, I mean some elemental fairness in the terms of work, life, and opportunity around the globe.

Philanthropy is generally harnessed to one of two great purposes: conservation of something of value, such as a forest or art; and constructive change in some human condition, behavior or system.

The philanthropy of conservation will always have appeal. But it is the second purpose—constructive change in some human condition, behavior or system—that is the most important area of philanthropic endeavor for this generation. Because of the ultimata we have discussed earlier, this may properly be called the philanthropy of human survival. And as we have already seen, the question of equity will be central to the philanthropy of survival.

How does that work?

Here's how. We are one planet now. Goods, services, weapons, pollution, disease, ideas, entertainment, drugs and people flow all over this globe without much regard to borders. The biggest problems we face—the threat of weapons of mass destruction, the environmental crunch, and the gap between the rich and the poor—cannot be solved on the scale of the nation-state. As we build the conceptual framework for looking at these problems within a global context, we will be drawn inevitably to formulate an elemental level of fairness, of opportunity, to which all people born on this planet have a right. It is a way of defining the social contract, the basic "deal." It's a way of saying: here's what your obligations as a citizen are, and here's what your basic rights are in return. At the root of a global social contract will be an idea of fairness, of equity. And very close to the practical center of that idea of equity will be access to minimal health care—for everyone. For every, single, living human being on this planet. We need foundations to fund the founding of the global social contract on this planet.

All Together: Equity, Science, and Accountability

And that brings us full circle—looking at equity, science and accountability together, in the whole. Can we get them together?

For me, the goal of equity is more important right now than the goal of better science. That doesn't mean that science is unimportant. Science is essential to meeting the goals of global minimum health care. But the most important work lies in building the expectations, the institutions, the financing and delivery mechanisms, and the structures that will allow affirmation and participation in health care by the individual in managing her own health at ground level. We need to do the simple things we already know how to do as well as develop new science to give populations tools to take better care of themselves.

And we need to spend money on this. Now. Health is the fastest and most obvious avenue through which to tackle the depressing and discouraging gap between the rich and the poor. And our country, the United States, needs to be a leader in this. A committed, morally motivated, and powerful leader . . . not a halting laggard that shortchanges world health while it finances militarization around the globe.

Our accountability in all this is not to some new body others might imagine to "oversee" foundations. Our accountability is not to the philanthropic titans who established these foundations; it is not to the past. Our accountability is to our children, and to our children's children. For they will see what we have done—or failed to do. It will not be difficult for them to judge us and our works. Indeed, it will be impossible for them not to.

So that is the dream and the agenda for the world of philanthropy in the area of health. Is it different from what the generation before us faced?

Yes.

First, it is one planet now in ways they could never have guessed.

Second, our sense of what lies ahead of us different. Rather than limitless vistas of progress and opportunity, we understand now that we have a relatively short window of time before an accelerating pattern of deterioration binds us all. Perhaps a very, very short window indeed.

Third, we know a lot about what can be done, and we know there is a lot we could put in place right now if we had the will and the money. And as a matter of fact—we do have the money. Where is the will?

Fourth, unlike the situation 25 years ago, there is a vast, vibrant and growing non-governmental organization sector around the world, consisting of resourceful and committed social entrepreneurs. The lofty, older and clunkier traditional institutions may be out of date. But there is a whole army of supple, adaptive, non-profit organizations out there all over the world that attract the youngest and the ablest, and who constitute an essential part of the infrastructure we shall need as we move forward with the health agenda.

Let's not keep them waiting too much longer.

So in all this should we be pessimistic?

No.

Hope.

Always hope.

In the words of the Chinese poet Lu Xun:

"Hope is like a path in the countryside. At first, there is no path. But then, as people are all the time coming and walking in the same way, a path appears."

Notes

1. *The Punjab, the Great North Plain of China, and the Great Plains of the North America.*

Index

www.ingramcontent.com/pod-product-compliance
Ingram Content Group UK Ltd.
Pitfield, Milton Keynes, MK11 3LW, UK
UKHW041142300926
15471UKWH00011B/97

9 780761 838197